PERGAMON INTERNATIONA
of Science, Technology, Engineering an(
The 1000-volume original paperback library i
industrial training and the enjoyment
Publisher: Robert Maxwell, M.C.

INTERNATIONAL SERIES ON
MATERIALS SCIENCE AND TECHNOLOGY
VOLUME 26—EDITOR: D. W. HOPKINS, M.Sc.

An Introduction to Chemical Metallurgy

SECOND EDITION
In SI/Metric Units

Other Titles in the International Series on
MATERIALS SCIENCE AND TECHNOLOGY

KUBASCHEWSKI, EVANS & ALCOCK
Metallurgical Thermochemistry, 4th Edition
RAYNOR
The Physical Metallurgy of Magnesium and its Alloys
PEARSON
A Handbook of Lattice Spacings and Structures of Metals and Alloys—Volume 2
MOORE
The Friction and Lubrication of Elastomers
WATERHOUSE
Fretting Corrosion
DAVIES
Calculations in Furnace Technology
REID
Deformation Geometry for Materials Scientists
BLAKELY
Introduction to the Properties of Crystal Surfaces
GRAY & MULLER
Engineering Calculations in Radiative Heat Transfer
MOORE
Principles and Applications of Tribology
GABE
Principles of Metal Surface Treatment and Protection
GILCHRIST
Extraction Metallurgy
SMALLMAN & ASHBEE
Modern Metallography
PARKER
An Introduction to Chemical Metallurgy, 2nd Edition
CHRISTIAN
The Theory of Transformations in Metals and Alloys, Part 1, 2nd Edition
HULL
Introduction to Dislocations, 2nd Edition
SCULLY
Fundamentals of Corrosion, 2nd Edition
SARKAR
Wear of Metals
HEARN
Mechanics of Materials
DOWSON & HIGGINSON
Elasto-hydrodynamic Lubrication, SI Edition

An Introduction to Chemical Metallurgy

SECOND EDITION

In SI/Metric Units

BY

R. H. PARKER, M.A., C.ENG., F.I.M.M., M.I.M.

Vice-Principal, Camborne School of Mines, England

PERGAMON PRESS

OXFORD · NEW YORK · TORONTO · SYDNEY

PARIS · FRANKFURT

U.K.	Pergamon Press Ltd., Headington Hill Hall, Oxford OX3 0BW, England
U.S.A.	Pergamon Press Inc., Maxwell House, Fairview Park, Elmsford, New York 10523, U.S.A.
CANADA	Pergamon of Canada Ltd., 75 The East Mall, Toronto, Ontario, Canada
AUSTRALIA	Pergamon Press (Aust.) Pty. Ltd., 19a Boundary Street, Rushcutters Bay, N.S.W. 2011, Australia
FRANCE	Pergamon Press SARL, 24 rue des Ecoles, 75240 Paris, Cedex 05, France
FEDERAL REPUBLIC OF GERMANY	Pergamon Press GmbH, 6242 Kronberg-Taunus, Pferdstrasse 1, Federal Republic of Germany

First edition 1967

Second edition 1978

British Library Cataloguing in Publication Data

Parker, Roger Hill
An introduction to chemical metallurgy.—2nd ed.
—(International series on materials science and technology; 26).
1. Chemistry, Metallurgical
I. Title. II. Series
669'.9 TN673 77-30266

ISBN 0-08-022125-4 (Hardcover)
ISBN 0-08-022126-2 (Flexicover)

Printed in Great Britain by Biddles Ltd., Guildford, Surrey

Contents

Preface

EARLY courses in metallurgy contained a large proportion of extraction metallurgy and metallurgical analysis, but the growth of physical metallurgy almost completely ousted these subjects from many courses, with the result that extraction metallurgy was often taught as a sideline with an inevitable concentration on practical details of processes rather than basic principles. The legacy of this period will be with us for some considerable time to come. The rise in importance of corrosion and surface treatment of metals, the realization that the physical metallurgist needs to know something of the applications of thermodynamics and reaction kinetics, and the development of experimental techniques with high-temperature systems has led to the resurgence of chemical metallurgy as a subject to be taught to metallurgists in its own right instead of as a fill-gap or as a potpourri of knowledge to be gleaned at odd points in a variety of lecture courses. The importance of a sound foundation in chemical metallurgy to the extraction metallurgist was well stated in the 16th Hatfield Memorial Lecture given to the Iron and Steel Institute by Professor F. D. Richardson in November 1964.

This book is written as an introduction to the subject and may be suitable for students embarking on courses for the Higher National Certificate, Institution of Metallurgists' examinations, Higher National Diplomas, or Degrees in Metallurgy. It is also hoped that it will help those entering the metallurgical field after training in another discipline, and those chemists who have to teach metallurgists, to appreciate the metallurgist's emphasis on certain points in chemistry which may be unfamiliar to them. A knowledge of chemistry, physics and mathematics of roughly G.C.E. A-level standard is assumed, and some topics in physics and inorganic chemistry, such as the kinetic theory of gases, atomic structure and the chemistry of the transition elements, should be studied in appropriate texts.

The early chapters in the book present the fundamental principles of the subject and the later chapters some of the applications of these principles. References in the text to a date and the author of some law or principle of physical chemistry are given for the sake of historical significance. Detailed references at the end of each chapter are those which it is considered are worthwhile following up for further reading, which represent more recent work of significance to the metallurgist, or which are an acknowledgement of sources of information and data. Readers looking for a rigorous treatment of each topic will have to look elsewhere, as this book is intended to present a point of view rather than act as a source book or as a fundamental treatise. It is hoped that some of the references for further reading will satisfy the need of the reader who wishes to inquire further into the history and details of the subject.

As far as possible, the symbols and abbreviations recommended in British Standard 1991: Part 1: 1954 (incorporating amendments issued since that date) have been used, as a uniform system of symbols is vital in this type of subject.

The author is indebted to all those—particularly students and colleagues at the Colleges of Technology in Rotherham and Sheffield—whose discussions have helped to make him feel that the writing of this book has been both necessary and worthwhile. He would also like to acknowledge those sources of information, diagrams and data which he has used—all of which are either mentioned in the text or in the references at the end of each chapter—and finally to thank Mrs. J. Vellenoweth for typing the manuscript and Mr. D. W. Hopkins for his encouragement and advice.

Preface to Second Edition

THIS edition has been revised to incorporate the use of the International System of Units (SI).

Symbols and Abbreviations

AS FAR as possible, the symbols and abbreviations recommended in British Standard 1991: Part 1: 1954, incorporating subsequent amendments, have been used.

A	frequency factor; area
a	Raoultian activity
C	heat capacity (usually molar heat capacity)
C_p	heat capacity at constant pressure
C_v	heat capacity at constant volume
c	concentration; specific heat
D	diffusion coefficient, diffusivity (molecular diffusion)
D_e	eddy diffusivity
E	activation energy; electrode potential; electromotive force
E_D	activation energy for diffusion
e	Henrian interaction coefficient/parameter
$\mathbf{F}$	Faraday's constant
F	force; work function, Helmholtz free energy
f	Henrian activity coefficient
G	Gibbs free energy
g	gravitational acceleration
H	enthalpy, heat content
$\mathbf{h}$	Planck's constant
h	height; Henrian activity
I	electric current
i	electric current density
J	mass flux

K	equilibrium constant
$\mathbf{k}$	Boltzmann's constant
k	velocity constant, rate constant, specific reaction rate
k_m	mass transfer coefficient
L_t	heat of transformation
L_e	latent heat of evaporation
L_f	latent heat of fusion
L_s	latent heat of sublimation
l	length
M	atomic weight; molecular weight
m	mass
$\mathbf{N}$	Avogadro's number
N	number of molecules
n	number of moles; order of reaction
$\mathbf{P}$	steric probability factor
P	pressure; total pressure of a system
p	partial pressure; vapour pressure
Q	electric charge
q	heat absorbed by system
q_p	heat absorbed by system at constant pressure
q_v	heat absorbed by system at constant volume
$\mathbf{R}$	gas constant
R	electrical resistance
r	radius; distance between electric charges
r_c	critical radius
S	entropy
T	absolute temperature
t	time; transport number
$t_{1/2}$	half-life
$T.P.$	throwing power
U	internal (intrinsic) energy
U_o	lattice energy
u	ionic mobility (velocity)

V	volume; potential difference
V_D	decomposition potential
v	velocity; (volume)
W	thermodynamic probability; work of adhesion or cohesion
w	work done by system
x	atom fraction; ion fraction; mole fraction
y	thickness of oxide film
Z	collision number
z	valency of an ion
α	degree of dissociation
Γ	excess surface concentration
γ	interfacial energy, interfacial tension; Raoultian activity coefficient; ratio of specific heats
δ	effective boundary layer thickness
$\in$	dielectric constant
ϵ^-	negative charge of one electron – used in equations for electrode reactions
ϵ	Raoultian interaction coefficient/parameter
η	overpotential
θ	contact angle
κ	specific conductance, conductivity
Λ	equivalent conductance
Λ_o	equivalent conductance at infinite dilution
Λ_{X^+}	equivalent conductance of the ion X^+
Λ_m	molar conductance
μ	chemical potential
Φ	fugacity
ρ	density; specific resistance, resistivity
ψ	roughness factor

Superscripts

^{o}	standard value of a thermodynamic property (ΔG^o)

	equilibrium value of a variable (c'_A)
ΔG^M	integral free energy of mixing (ΔH^M, ΔS^M, etc.)
$\bar{}$	partial molar quantity ($\bar{G}_A$)
^{E}	excess molar quantity ($\bar{G}_A^E$)
^{I}	ideal molar quantity (G^I)
*	thermodynamic variable associated with the formation of the activated complex of a reaction (ΔG^*)

Subscripts

G, L or S	accompanying chemical formulae refer to the physical state of a substance; gaseous, liquid or solid respectively
aq	accompanying a chemical formula indicates that the substance is in aqueous solution

$\ln x$	natural logarithm of x
$\log x$	logarithm to the base 10 of x
°C	degrees Celsius
°K	degrees Kelvin (absolute scale of temperature)
J	joule(s)
e.m.f.	electromotive force
cal	calorie(s)
kcal	kilocalorie(s)
V	volt(s)
A	ampère(s)
g	gramme(s)
Å	ångström(s)
mole	gramme-molecule
( ) or [ ]	surrounding chemical formulae referring to slag/metal reactions denote components of the slag phase and the metal phase respectively

CHAPTER 1

Introduction to Thermodynamics

1.1. Introduction

A dictionary defines thermodynamics as "the science of the relations between heat and mechanical work", but it might more usefully be defined as the study of the changes in energy accompanying chemical and physical changes, which allows experimentally determined laws to be derived from certain basic principles, and helps to predict changes which have not been observed. It must be remembered that thermodynamics considers only the initial and final states of any system undergoing a change, and provides no information about the mechanism of the change between these states, or the rate at which the change takes place—this is the subject of reaction kinetics (where the change taking place is a chemical reaction), and will be considered in Chapter 4.

Rumford, in the late eighteenth century, showed that mechanical work could be changed into heat, and Joule, in the middle of the nineteenth century, established the "mechanical equivalent of heat"—the observation that there was a definite relationship between the amount of heat energy and the mechanical work which produced it. Since then many workers have contributed to the science of thermodynamics—mathematicians, physicists and chemists—but it is only comparatively recently, since satisfactory experimental techniques at high temperatures have been developed, that thermodynamics has been successfully exploited in the metallurgical field. Developments and improvements in methods of metal extraction and refining have been a direct result, and its application has produced a more fundamental understanding of phase diagrams, corrosion, and physical phenomena such as precipitation in alloys.

This introduction to thermodynamics is intended for the metallurgist who is either meeting the subject for the first time, or has passed through the treatment given by a chemist, physicist or engineer—none of whom consider it quite from the viewpoint of the metallurgist. It is probably true to say that metallurgical thermodynamics is more closely allied to chemistry than to any other discipline. Readers who wish to consult other works on the subject will find a list of references at the end of this chapter.[1,2]

1.2. Energy

The *energy* of a body can be defined as its capacity for doing work, based on the fact that this energy can be translated into mechanical work, which can be measured. We have already seen that mechanical work can be transformed into heat, but there are many other forms of energy, such as electrical, chemical and surface energy, potential and kinetic energy, all of which are interchangeable.

The work done when a force of 1 newton acts through a distance of 1 metre is the unit of energy known as the *joule*. (The *newton* is the force which produces a velocity of 1 m/s when acting upon a mass of 1 kg for 1 second.) Many text books and papers use a unit of thermal energy which is not recognized in the SI system of units. This is the *calorie*, which is equivalent to 4·184 joules, and is the amount of energy required to raise the temperature of 1 g of water from 14·5 to 15·5°C.

1.3. The First Law of Thermodynamics

This is a law based on experience, and is an extension of the Principle of Conservation of Energy which states that energy cannot be created or destroyed, provided that there is no measurable conversion of mass to energy; it can only be converted from one form to another. The consistency of Joule's mechanical equivalent of heat is explained by this principle, and it is clear that if it is obeyed, no machine can be devised which will do work without the expenditure of an equivalent amount of energy. In order to melt a given mass of a metal, a certain amount of energy must be supplied—whether by combustion of a fuel or the expenditure of electrical energy.

Thermodynamics specifies a "system" as any matter which is being considered which consists of a definite amount of a given substance or substances. The system will have its "surroundings" with which it can exchange energy, and the system and its surroundings are considered to be *isolated*—they cannot exchange energy with any other system. An example of such a system would be a steel billet in a reheating furnace—the surroundings of the billet would include not only the structure of the furnace itself, but also the atmosphere inside and outside the furnace, the ground on which the furnace stood, and any other objects which could exchange a measurable amount of energy with the billet, either directly or indirectly.

Thus, the First Law of Thermodynamics states that the total energy of a system and its surroundings remains constant, even if it may be changed from one form of energy to another.

We can now introduce a quantity U, which represents the total energy of a system—whether it be kinetic, electrical, rotational, vibrational, or any other form of energy except the energy due to its position in space, which is assumed to be constant. U is called the *internal* or *intrinsic* energy of the system and is not usually known quantitatively because of the difficulty of measuring all the different forms of energy possessed by actual systems. This is not important because we are concerned with *changes* in energy, which can be measured. If the state of a system changes from A to B, for instance as a result of a change in temperature, then we say that the change in internal energy is the internal energy in the final state, U_B, minus the internal energy in the initial state, U_A, or

$$\Delta U = U_B - U_A, \tag{1.1}$$

where ΔU represents a *finite* change in U. We have thus introduced a **sign convention**, because, if the internal energy of the system in the final state is greater than in the initial state, ΔU is positive, and if the system loses energy, ΔU is negative. This sign convention must be adhered to throughout the application of thermodynamics to avoid serious confusion. A common error is to consider that because something *feels* warm to the observer (who is part of the *surroundings*, not of the system being considered), then the heat change is positive because heat is being produced. The sign of an energy change in a system must be considered from the point of view of the *system*, not of its surroundings, and any loss of energy from the system, such as that causing a sensation of warmth to an observer, must be represented by a negative sign. A freezing mixture employs some form of reaction—such as the solution of a salt—which absorbs heat energy, and this is represented by a positive change in energy with respect to the freezing mixture system, and according to the First Law of Thermodynamics an equal change of energy, but of opposite sign, with respect to the surroundings, which are cooled down.

The First Law of Thermodynamics can be expressed mathematically by considering a change in a system involving only two forms of energy change—heat and mechanical work. For example, if a gas is heated, it gains heat energy, and probably does mechanical work by expanding against an external pressure. Alternatively, if the gas is cooled, it may contract, with the result that external work is done on it, and there will be a loss of heat energy as a result of the cooling.

If we represent the work done *by* the system as w, and the heat absorbed *by* the system as q,

$$\Delta U = q - w \tag{1.2}$$

by the First Law of Thermodynamics, and represents the internal energy change of a system resulting from changes in heat and mechanical energy. In applying equations like (1.2), care must be taken that the energy terms are all expressed in the same units. If ΔU is to be expressed in calories, the units of w must be changed from mechanical to heat energy units by the use of the mechanical equivalent of heat. If the SI system of units is used, the joule is the unit used for all forms of energy, and no problem will be encountered in this respect.

1.4. The Expansion of a Gas: Thermodynamic Variables

The work done in a given physical change can be measured and is illustrated by considering the simple example of an ideal gas, which obeys the Ideal Gas Equation (see ref. 1, p. 193)

$$PV = \boldsymbol{R}T, \tag{1.3}$$

where P is the pressure in N/m^2, V the volume in m^3 occupied by one mole, $\boldsymbol{R}$ the gas constant and T the absolute temperature in degrees K. One mole is the amount of substance of a system which contains as many elementary entities (atoms, molecules, ions, electrons, etc.) as there are atoms in ·012 kg of the isotope carbon 12. The value of $\boldsymbol{R}$ is 8·314 J/°K mol.

Figure 1.1 shows this gas enclosed in a cylinder by a weightless, frictionless piston—these limitations allow us to neglect energy changes which would be the result of friction between piston and cylinder and movement of the mass of the piston. Assuming that the cross-sectional area of the cylinder is x square metres, the force on the piston is Px newtons. If the gas expands so that the piston moves y metres, then the work done by the gas against the pressure

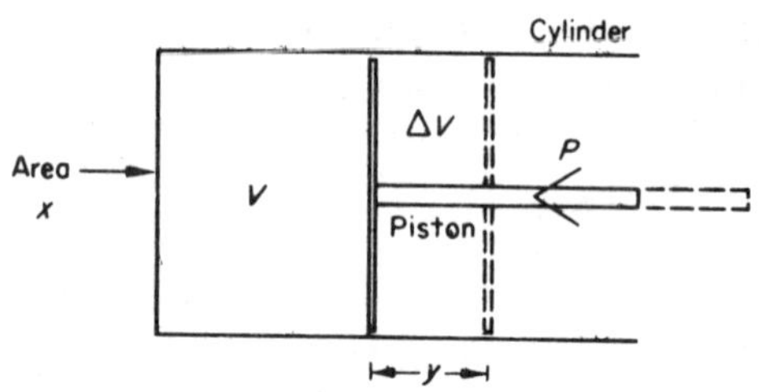

FIG. 1.1. Expansion of a gas, initially at volume V, by ΔV against pressure P in a cylinder of cross-sectional area x. The piston enclosing the gas in the cylinder moves a distance y.

on the piston is Pxy joules. But xy is ΔV, the change in volume of the gas, and therefore,

$$w = P\Delta V. \tag{1.4}$$

This is provided that the pressure remains constant throughout the expansion, and can be expressed by plotting a graph of pressure against volume (Fig. 1.2), where the work done in

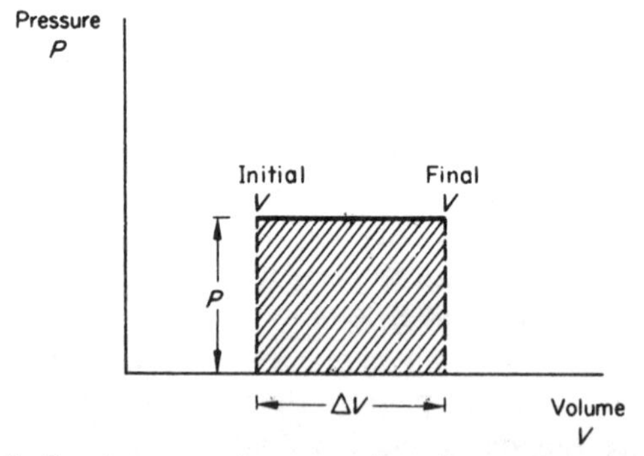

FIG. 1.2. Graph representing expansion of a gas by a volume ΔV against constant pressure P. The work done in the expansion is represented by the shaded area.

the expansion is represented by the shaded area under the graph.

If the pressure does not remain constant throughout the expansion, then the work done is obtained by integrating the expression PdV, where the integration represents the summing of each increment of work done as the volume changes by a series of infinitesimally small amounts dV (Fig. 1.3). Again,

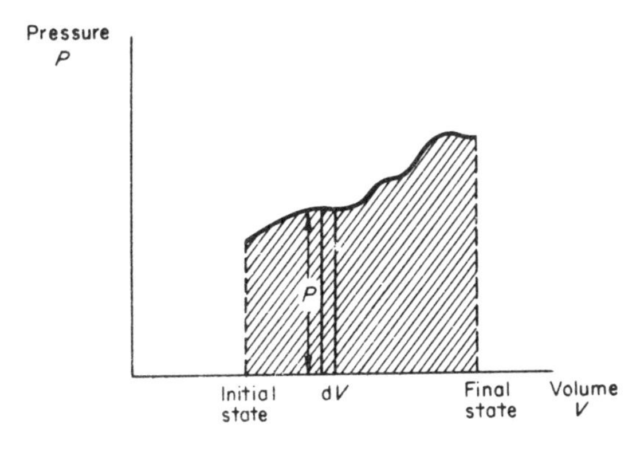

FIG. 1.3. Graph representing expansion of a gas against a varying pressure P. The work done in the expansion is represented by the shaded area—the sum of area increments $P \cdot dV$.

the work done is represented by the shaded area under the curve,

$$w = \int P dV. \tag{1.5}$$

Using (1.2), at constant pressure

$$\Delta U = q_p - P\Delta V, \tag{1.6}$$

where q_p is the heat absorbed by the system at constant pressure.

At constant volume, no external work is done, $\Delta V = 0$, and

$$\Delta U = q_v. \tag{1.7}$$

If a system changes from state A to state B, the total internal energy change will be the same, ΔU, where, according to (1.1), $\Delta U = U_B - U_A$. This must apply, whatever the method employed in changing from A to B, in order that the First Law of Thermodynamics be obeyed.

Figure 1.4 demonstrates this—where the state of the gas in the cylinder of Fig. 1.1 is changed from A to B by I moving the piston against constant pressure, II against a varying pressure. In case I, the volume is increased by $\Delta V = V_B - V_A$

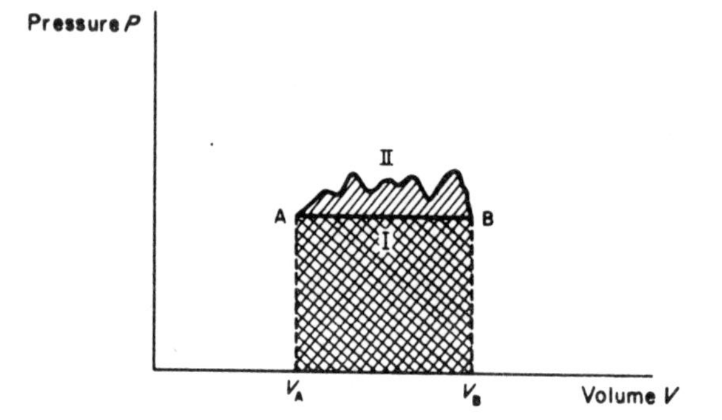

FIG. 1.4. Graph representing expansion of a gas by two methods: I at constant pressure; II at a varying pressure. V_A and V_B are the initial and final volumes respectively. The work done in expansion I is shown by the cross-hatched area. The work done in expansion II exceeds that in expansion I by the singly shaded area.

by a simple, direct route—in case II, the same overall volume change is achieved by a complicated, indirect route. ΔU will be the same in each case by the First Law of Thermodynamics, and we call a variable such as U a *thermodynamic variable*—a change in which is independent of the path taken, depending only on the initial and final states of the system. U is a property of the system, independent of its history, and other thermodynamic variables which we have already mentioned are temperature, pressure and volume—none of which depends on the history of a system, only on its existing state. (An alternative to the term "thermodynamic variable" is "property of state"—that is a property used to describe the state of a system.)

If we consider the work done by the gas in the expansions represented by Fig. 1.4, we will see that in case I, $w_{\text{I}} = P\Delta V$, which is much less than $w_{\text{II}} = \int_{V_A}^{V_B} P\,dV$, as shown by the areas under the two curves I and II. It will be clear that the work done by the system is not independent of the path taken during a change in the system, and w is not, therefore, a thermodynamic variable.

From (1.2), $\Delta U = q - w$, so that if ΔU is to remain the same whatever the path taken between A and B, q must vary according to the path taken just sufficiently to compensate for the corresponding variations in w, and vice versa. q is therefore not a thermodynamic variable.

Mathematically, thermodynamic variables have complete differentials, dU, dP, etc., whereas variables such as q and w do not have complete differentials, and we refer to infinitesimal changes in these variables as incomplete differentials, δq, δw, etc.

A further distinction between variables must be explained at this stage. Certain properties of a system—such as temperature, pressure, surface tension, density, refractive index, viscosity—do not depend on the mass of the system and are called *intensive* properties. Other properties—such as volume and internal energy—are dependent on the amount of the substance or substances present in the system, and are called *extensive* properties.

1.5. Thermodynamically Reversible Changes

Considering a cylinder and piston of the type shown in Fig. 1.1; we can expand the gas in the cylinder in two ways. If we carry out the expansion so that no property of the system ever differs by more than an infinitesimal amount from one instant in time to the next, by reversing this process, the gas can be returned to its original state without any overall change in either the system or its surroundings, and such a process is "*thermodynamically reversible*". The system at all times remains in equilibrium with its surroundings and the process is necessarily infinitely slow. If the expansion is carried out so that a property of the system differs by a finite amount from one instant to the next, the system cannot return to its original state without some changes occurring in the surroundings—and such process is *thermodynamically irreversible*. (Chemical reactions can be made to proceed in the forward and reverse directions, but a chemical change is only *thermodynamically reversible* if at all times the system remains in equilibrium, and the change is infinitely slow.) A rapid movement of the piston resulting in a finite decrease in gas pressure would cause a drop in the temperature of the gas, heat would be transferred from the surroundings, and, in order for the process to be reversed completely, heat would then have to pass out from the gas to its surroundings which would then be at a higher temperature than the gas—an impossible state of affairs without some external cooling of the surroundings, which would be a permanent change in their properties. It will be seen that a reversible process will take an infinite time to complete, and therefore reversible processes are hypothetical and can never be achieved—natural processes must be irreversible.

In the case of the gas expansion, if the gas is expanding against a pressure P, with a pressure $P+\mathrm{d}P$, the process is reversible, whereas if its pressure is $P+\Delta P$, the process is irreversible. By reducing ΔP to a very small interval of pressure, the gas can be allowed to expand under conditions approximating to reversibility, within the experimental error of the measurements involved. Certain other processes such as the chemical reaction in an electrical cell and the evaporation and condensation of liquids and gases can be made almost reversible under carefully controlled conditions, but the vast majority of processes—physical and chemical—must be irreversible. An example of the reaction in an electrical cell behaving almost reversibly is the reaction which takes place in the Daniell cell—which consists of a zinc electrode dipping into zinc sulphate solution separated by a porous pot from a solution of copper sulphate into which a copper electrode is dipped. If the two electrodes are connected by an electrical conductor, a current flows through the conductor because of the tendency of the following reactions to take place, as will be discussed in Chapter 5,

$$Zn = Zn^{2+} + 2\epsilon^-$$

$$Cu^{2+} + 2\epsilon^- = Cu.$$

The electrons released at the zinc electrode flow through the conductor to the copper electrode where they combine with the copper ions in the solution to allow the deposition of copper on the electrode. The overall cell reaction is therefore

$$Zn + Cu^{2+} = Zn^{2+} + Cu.$$

If the Daniell cell is opposed by a variable e.m.f., this can be controlled so that it almost exactly balances the e.m.f. of the Daniell cell, and no measurable current flows in the circuit. If the variable e.m.f. is decreased slightly, the above reaction in the Daniell cell proceeds from left to right, and if the variable e.m.f. is increased slightly, the reaction is reversed and zinc will plate out and copper dissolve. The reaction in the cell is then behaving thermodynamically reversibly within the margin of experimental error. If the difference in e.m.f. between the Daniell cell and the opposing e.m.f. is large, "polarization" will occur as a result of changes in concentration of the solutions around the electrodes, and there may even be heat losses due to the heavy currents passing through the conductors and the cell itself. The cell would then be behaving irreversibly, as an equal and opposite applied e.m.f. would not then completely reverse the cell reaction—for example, the heat lost in the forward reaction would not be recovered in the backward reaction.

This concept of thermodynamically reversible processes will be expanded in Chapter 2, after consideration of the implications of the Second Law of Thermodynamics, but it is important to realize here that by initially considering reversible processes, always in equilibrium, we can express energy changes in a system in terms of properties such as pressure, temperature, volume and other thermodynamic variables which can be measured experimentally. Knowledge of equilibrium conditions is necessary for the full understanding of non-equilibrium conditions in a system.

1.6. "Heat Content" or "Enthalpy"

From (1.6), at constant pressure,

$$\Delta U = q_p - P\Delta V.$$

Therefore

$$\begin{aligned} q_p &= \Delta U + P\Delta V \\ &= (U_B - U_A) + P(V_B - V_A) \\ &= (U_B + PV_B) - (U_A + PV_A), \end{aligned}$$

where A and B are the initial and final states of the system respectively.

We can now introduce a new thermodynamic variable—an extensive property of the system called *heat content* or *enthalpy*, H, so that

$$H = U + PV. \qquad (1.8)$$

Then

$$H_A = U_A + P_A V_A$$

$$H_B = U_B + P_B V_B,$$

and at constant pressure, $P = P_A = P_B$, so that

$$q_p = H_B - H_A = \Delta H.$$

This means that the increase in enthalpy of the system, ΔH, accompanying a change in the system at constant pressure is equal to the heat absorbed by the system, q_p; hence the alternative name "heat content" for the variable H.

$$\Delta H = \Delta U + P\Delta V, \tag{1.9}$$

at constant pressure, and, as U, P and V are all thermodynamic variables, H must also be a thermodynamic variable and is independent of the previous history of the system—depending only on its present state. In other words, it is a "property of state".

The heat absorbed or evolved in a chemical reaction (usually carried out at constant pressure in metallurgical processes) is an important property of the reaction, which can be measured and is the basis of the subject of thermochemistry which will be discussed in Section 1.9.

1.7. Gas Expansion: Maximum Work

We have already seen (Section 1.4) that the work done by a gas in expanding at constant pressure is

$$w = P\Delta V,$$

but this is only the work produced by the thermodynamically reversible expansion of the gas. If the expansion was irreversible, the pressure against which the gas expanded would be less than P, the pressure inside the cylinder, and therefore the work done would be less than $P\Delta V$, the maximum work done as a result of the expansion. The work done by reversible processes is always greater than that done by the same processes carried out irreversibly, and is therefore called the *maximum work* for the processes.

From (1.5) when P is not constant,

$$w = \int_{V_A}^{V_B} P \mathrm{d}V,$$

but $PV = RT$ for 1 mole of ideal gas, so that

$$w = \int_{V_A}^{V_B} \frac{RT}{V} \mathrm{d}V = RT \ln \frac{V_B}{V_A} \tag{1.10}$$

provided that the expansion is "isothermal" (at constant temperature).

But $P_A V_A = P_B V_B$ (from the ideal gas equation) so that

$$w = RT \ln \frac{P_A}{P_B}. \tag{1.11}$$

We have thus obtained expressions for the work done in the reversible expansion of a gas at constant pressure and at constant temperature, and know that the work done in the irreversible expansion of a gas is always less than these expressions for maximum work.

1.8. Heat Capacity: The Temperature-dependence of Enthalpy Changes

Different substances require different amounts of heat energy in order for their temperature to be raised by a given interval—for instance, one gramme of mild steel requires 0·460 J to raise its temperature by 1°C, whereas one gramme of copper requires 0·385 J; these figures can be contrasted with 0·795 for firebrick and silica brick and approximately 0·858 for a moulding sand. If the temperature of the material is raised to 1000°C, these values can rise by as much as 50%. This clearly is a property of a substance which is important—for instance, in the calculation of thermal requirements for heating metal ingots and billets, or of the chilling effect of certain moulding materials.

We define the *specific heat* of a substance, c, as the amount of heat required to raise the temperature of 1 g of that substance by 1 degC—and the temperature at which the appropriate measurement was made must be stated, as c generally varies with temperature.

In thermodynamics, we are concerned with "systems"—and we define the *heat capacity* of a system, C, as the amount of heat required to raise the temperature of that system by 1 degC—again stating the temperature. We have found (Section 1.4) that 1 mole is a useful quantity of a system to consider, so that usually the heat capacity of 1 mole of a system—the *molar heat capacity*—is implied in thermodynamic expressions, and will be used in this book.

From the definition of heat capacity,

$$C = \frac{\delta q}{\mathrm{d}T}$$

the rate of change of heat absorbed by the system with increase in temperature. The heat capacity at constant pressure, C_p, is therefore

$$C_p = \frac{\delta q_p}{\mathrm{d}T},$$

but $q_p = \Delta H$ (Section 1.6 above), so that

$$C_p = \left(\frac{\partial H}{\partial T}\right)_p, \tag{1.13}$$

where the partial differential is used to allow for the fact that H depends on both pressure and volume, and here the pressure is kept constant. Similarly, the heat capacity at constant volume

$$C_v = \frac{\delta q_v}{\mathrm{d}T} = \left(\frac{\partial U}{\partial T}\right)_v \tag{1.14}$$

because, from (1.7), $q_v = \Delta U$.

Heat capacities depend on temperature, and, although theories have been produced to explain this dependence,[3] the results of experimental determinations are expressed in empirical relationships such as

$$C_p = \mathrm{a} + \mathrm{b}T + \mathrm{c}T^{-2}, \tag{1.15}$$

where a, b and c are constants determined for a substance over a particular range of temperatures. These constants are listed in tables such as those provided by Kubaschewski and Evans.[5] Curves of the variation of heat capacities with temperature can be plotted, and an example is shown in Fig. 1.5. Here, nickel shows a gradually increasing heat capacity with temperature between 0 and 900°K, except for a sharp discontinuity at about 600°K. This corresponds to the magnetic

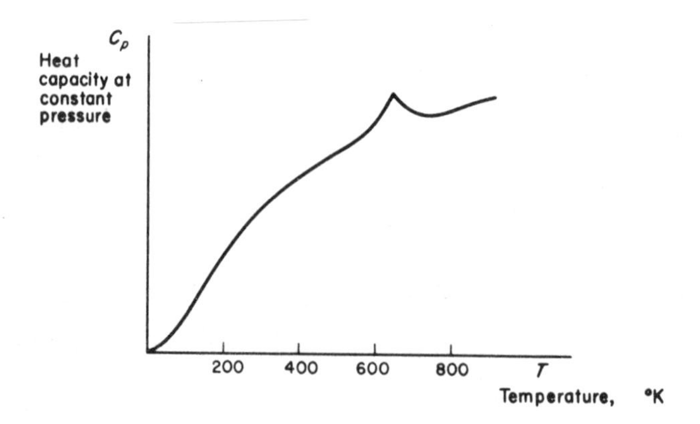

FIG. 1.5. Atomic heat capacity of nickel from 0° to 900°K. (After Kubaschewski and Evans.[5])

transformation of nickel. Discontinuities in these curves occur wherever there is a phase change – such as the α to γ change in pure iron and the melting of a solid – and also when there are order–disorder transformations in solid alloys.

From (1.13), enthalpy changes are related to heat capacities, so that enthalpy changes would be expected to be temperature-dependent. If we consider the enthalpy change accompanying the change of a system from state A to state B, then

$$\Delta H = H_{\mathrm{B}} - H_{\mathrm{A}}.$$

Differentiating with respect to temperature, keeping the pressure constant,

$$\left[\frac{\partial(\Delta H)}{\partial T}\right]_p = \left(\frac{\partial H_{\mathrm{B}}}{\partial T}\right)_p - \left(\frac{\partial H_{\mathrm{A}}}{\partial T}\right)_p.$$

But $(\partial H/\partial T)_p = C_p$ [from (1.13)], so that

$$\left[\frac{\partial(\Delta H)}{\partial T}\right]_p = C_{p(\mathrm{B})} - C_{p(\mathrm{A})} = \Delta C_p, \qquad (1.16)$$

where ΔC_p is the difference between the heat capacity of the system in its final state and the heat capacity of the system in its initial state at constant pressure. This is one of Kirchhoff's equations (G. R. Kirchhoff, 1858), and the other can be derived similarly, so that

$$\left[\frac{\partial(\Delta U)}{\partial T}\right]_v = \Delta C_v. \qquad (1.17)$$

Equation (1.16) is useful to the metallurgist, especially in its integrated form, because he is interested in high-temperature conditions; integrating,

$$\int_{\Delta H_1}^{\Delta H_2} \mathrm{d}(\Delta H) = \Delta H_2 - \Delta H_1 = \int_{T_1}^{T_2} \Delta C_p \,.\, \mathrm{d}T, \qquad (1.18)$$

where ΔH_2 and ΔH_1 are the enthalpy changes at temperatures T_2 and T_1 respectively.

If C_p is not constant for the system over the range of temperatures concerned, then the empirical expression of the type shown in (1.15) is inserted in (1.18):

$$\Delta H_2 - \Delta H_1 = \int_{T_1}^{T_2} (\mathrm{A} + \mathrm{B}T + \mathrm{C}T^{-2}) \, . \, \mathrm{d}T.$$

$$= (\mathrm{A}T_2 + \tfrac{1}{2}\mathrm{B}T_2^2 - \mathrm{C}T_2^{-1}) - (\mathrm{A}T_1 + \tfrac{1}{2}\mathrm{B}T_1^2 - \mathrm{C}T_1^{-1}).$$

Thus, if ΔH_1 is known – say in the case of the heat evolved in a chemical reaction – ΔH_2 can be calculated from the known values of C_p. An example of the application of Kirchhoff's equations to chemical reactions will be considered in detail in the next section.

When no heat energy enters or leaves the system during a change in the system (for example, when the system is placed in a well-insulated chamber), the change is known as *adiabatic*. In this case, the term q in (1.2) is zero, and the temperature will not remain constant.

It is found that, assuming that, for 1 mole, the difference between C_p and C_v is $\boldsymbol{R}$, the gas constant,

$$P_1V_1^\gamma = P_2V_2^\gamma \tag{1.19}$$

when the system changes adiabatically and reversibly from P_1, V_1 and T_1 to P_2, V_2 and T_2 respectively. γ is the ratio C_p/C_v for an ideal gas and is greater than unity. For a full proof of this relationship, readers should consult ref. 1, pp. 194 and 198.

1.9. Thermochemistry and its Applications in Metallurgy

In the casting of metals, the top of the mould can be constructed of substances which evolve heat when they come into contact with a source of heat, such as molten metal. This evolution of heat by the "exothermic" materials, as they are often called, keeps the metal at the top of the casting molten, allowing it to "feed" into the casting and fill the contraction cavity which forms when the metal solidifies. One of the reactions taking place in the exothermic material is the combustion of aluminium powder,

$$2Al+\tfrac{3}{2}O_2=Al_2O_3, \qquad \Delta H=-1{,}674{,}000 \text{ J/mol.}$$

The heat change accompanying this reaction is large and *negative*—so that a large quantity of heat is evolved and the reaction is known as an *exothermic* reaction—hence the name of the materials used in the top of the mould to produce heat.

Carbon is used to reduce zinc oxide in the commercial extraction of zinc and the reaction can be represented as

$$ZnO+C=Zn+CO, \qquad \Delta H=+349{,}000 \text{ J/mol.}$$

Because this reaction absorbs a large quantity of heat energy, the energy is supplied by external heating of the retorts in which the reaction is carried out, with a resultant expensive consumption of fuel. Reactions which absorb heat are called *endothermic* reactions, and the heat energy change accompanying the reactions will be *positive*.

The measurement of the heat energy produced or absorbed by a reaction is of theoretical interest as well as of commercial importance; the amount of heat evolved or absorbed in the formation of a compound tells us something of the structure of that compound in relation to those of the substances from which it was formed, and *thermochemistry*—the study of heat changes in chemical reactions—has made an important contribution to the understanding of metallurgical phenomena.

Metallurgical processes are usually carried out at constant pressure, so that we will consider the heat evolved by the system at constant pressure q_p, which we have seen (Section 1.6) is the change in *enthalpy* ΔH. In this case the "system" initially consists of the reactants and, finally, the products of the reaction. Enthalpy is an extensive property of the system—depending on the amounts of substances present in the system—so that we must be careful to state clearly the amounts of substances involved in the reaction. If the reaction is exothermic, we have seen that ΔH must have a negative sign, and if it is endothermic ΔH must be positive, using the sign convention in Section 1.3. As enthalpy changes are temperature-dependent, the temperature must be stated for every value of ΔH quoted.

Definitions of the enthalpy changes involved in different types of reaction are in common use, and each clearly states the amounts of reacting substances as follows:

Heat of Reaction: the change in enthalpy when the amounts of reactants in moles, shown by the balanced equation of a reaction, react completely as shown in the equation.

An example of the heat of a reaction is, at 727°C,

$$3Fe_2O_3+CO=2Fe_3O_4+CO_2, \qquad \Delta H=-46{,}700 \text{ J/mol.}$$

This is the heat evolved at constant pressure when 3 moles of haematite are reduced by 1 mole of carbon monoxide to form 2 moles of magnetite and 1 mole of carbon dioxide, as might occur in the iron blast furnace. Heats of reaction are often large and the unit of energy used for convenience may be the kilojoule (1000 J), so that in this case, ΔH would be −46·7 kJ. It is important to indicate the physical states of the reactants and products, as this will affect the value of ΔH. Thus,

$$3Fe_2O_{3_S}+CO_G=2Fe_3O_{4_S}+CO_{2_G}, \qquad \Delta H=-46{\cdot}7 \text{ kJ}$$

S signifies the solid, L the liquid, and G the gaseous states; where different allotropic modifications exist—as in the case of α-, γ- and δ-iron—these may also be indicated where necessary.

Heat of Formation of a Compound: the change in enthalpy when 1 mole of the compound is formed from its constituent elements in their stable forms at 25°C and 1 atm pressure. For example, the heat of formation of lead sulphide,

$$Pb_S + \tfrac{1}{2}S_{2_S} = PbS_S, \qquad \Delta H_{298} = -94{\cdot}1 \text{ kJ};$$

the subscript 298 refers to the temperature of the measurement, 298°K.

Since the absolute value of the heat content of a substance is not known, it has become the practice in thermochemistry to consider the heat content of elements in their *standard states*—at equilibrium at 25°C and 1 atm pressure—to be zero. This arbitrary zero of heat content of substances has no fundamental physical meaning, but as thermochemistry is concerned with *changes* in heat content accompanying reactions, there is no obstacle to the adoption of this convention. On this basis, the *heat of formation* of a compound as defined above is the *heat content* of that compound at 25°C. In the above example, the heat contents of the lead and the sulphur are considered to be zero, and the heat content of the lead sulphide is $-94{\cdot}1$kJ.

$$\Delta H_{298} = -94{\cdot}1 - (0+0).$$

Taking another example, the roasting of lead sulphide to lead oxide,

$$\begin{array}{cccccccl} 2PbS & + & 3O_2 & = & 2PbO & + & 2SO_2, & \Delta H_{298} = -845{\cdot}2 \text{ kJ.} \\ 2(-94{\cdot}1) & & 0 & & 2x & & 2(-297{\cdot}1). & \end{array}$$

Given that the heat of reaction at 298°K is $-845{\cdot}2$ kJ, calculate the heat content at 298°K of PbO. Let x kilojoules be the unknown heat content.

$$\Delta H_{298} = -845{\cdot}2 = 2x - 594{\cdot}1 - (-188{\cdot}2 + 0)$$
$$= 2x - 594{\cdot}1 + 188{\cdot}2$$
$$\therefore x = -219{\cdot}7 \text{ kJ/mol of PbO.}$$

Values of heats of formation are tabulated in books of reference according to these principles.[5]

Heat of Combustion of a Substance: the enthalpy change when 1 mole of the substance (an element or a compound) is completely burnt in oxygen.
For example,

$$Mg + \tfrac{1}{2}O_2 = MgO, \qquad \Delta H_{298} = -602{\cdot}5 \text{ kJ},$$
$$C + O_2 = CO_2, \qquad \Delta H_{298} = -393{\cdot}7 \text{ kJ}.$$

It will be seen that these values for the heats of combustion of magnesium and carbon are also heats of formation of their respective oxides, and, as combustion experiments can be carried out satisfactorily in a bomb calorimeter (see Section 1.10), measurements of heats of combustion supply a good starting point in the accumulation of thermochemical data. The heat of combustion of a fuel is the same as its "calorific value", and has industrial importance as a property by which the fuel may be evaluated.

Heat of Transformation: the change in enthalpy when 1 mole of a substance undergoes a specific physical change (such as melting, evaporation, allotropic modification). This general term can be given the symbol L_t, but is commonly described according to the transformation taking place. For example, the heat of transformation when zinc changes from solid to liquid at 420°C is 7·1 kJ/mol, and is called the *latent heat of fusion* L_f of zinc. This is the heat required to transform the crystalline structure of the solid metal into the less ordered structure of the liquid. The *latent heat of evaporation* L_e of zinc at 907°C is 114·2 kJ/mol, and this difference in magnitude between latent heats of evaporation and latent heats of fusion in metals reflects greater difference in structure between liquid and gas compared with the difference between solid and liquid. In the extraction of zinc, the metal can be produced as a vapour, which has to be condensed before it is cast as a liquid into moulds, and these values for the heat changes involved in condensation and solidification are important.

(Latent heats of fusion and evaporation are often quoted in calories per gram, rather than kilocalories per mole, and care should be taken to note the units of figures quoted in tables of reference. The units used here are consistent with the units used in the thermodynamic arguments presented in this book.)

When titanium transforms in the solid state from α (close-packed hexagonal) to β (body-centred cubic) at 880°C, the heat of transformation L_t is 3·4 kJ/mol. Allotropic transformations occur in iron, manganese, tin, uranium and other metals, and are of importance in the study of the thermodynamics of alloy systems. Enthalpy changes also occur when order–disorder transformations take place in solid alloys (e.g. the gold–copper system), and when there are changes in magnetic properties (e.g. the Curie point, 760°C, at which iron loses its ferromagnetic properties and requires about 2·9 kJ/mol to upset the ferromagnetic domains). Some idea of the magnitude of these heats of transformation may be given by a comparison with 5·9 kJ/mol and 41·0 kJ/mol for the latent heats of fusion and evaporation of water at its melting and boiling points respectively.

Heat of Solution: when one substance *dissolves* in another there will be a change in enthalpy. This is called the heat of solution, and depends on the concentration of the solution, which should be stated. The heat of solution increases as the concentration of the solution decreases, and it is usual to quote the heat of solution as the enthalpy change when 1 mole of solute is added to form a solution of a particular concentration, or alternatively added to such a large volume of solution of a particular composition that no significant change in the composition of the solution occurs. The former definition is the "integral" heat of solution, the latter the "partial" or "differential" heat of solution.

The integral heat of solution of silicon in liquid iron at 1580°C to form a 0·1 atom fraction of silicon in iron is −11·7 kJ mol – this corresponds to a steel containing approximately 4% silicon by weight – and it is a feature of the production of these steels (used for transformer laminations) that a rise in temperature of about 50°C is experienced when sufficient silicon is added to the molten steel in the ladle to produce a silicon content of 4% by weight. This rise in temperature of the steel must be taken into account when the steel is tapped into the ladle from the furnace, to prevent severe refractory attack because the steel is at too high a temperature after the silicon addition (see ref. 6, p. 816).

The First Law of Thermodynamics (Section 1.3) states that the total energy of a system and its surroundings remains constant, and as enthalpy is a thermodynamic variable (Section 1.4), the enthalpy change accompanying a reaction depends only on the initial and final states – not on the path taken. This is the basis of the experimental discovery of *Hess's Law of Constant Heat Summation* (1840), which states that the overall heat change for a chemical reaction is the same whether it takes place in one or several stages, provided the temperature and either the pressure or the volume remain constant.

This means that, to take an example, the heat of reaction for the reduction of haematite (Fe_2O_3) in the iron blast furnace might be calculated as follows:

Oxygen is blown into the furnace, and carbon is charged with the haematite, the products being carbon dioxide and iron (say). Then $Fe_2O_3 + 3C + \frac{3}{2}O_2 = 2Fe + 3CO_2$ could be the overall reaction whose heat of reaction we are to calculate. This reaction might take place in several stages – the combustion of carbon to form carbon dioxide, the reduction of the carbon dioxide by more carbon to form carbon monoxide, the reduction of haematite to magnetite (Fe_3O_4) by carbon monoxide, and finally the reduction of magnetite to iron—

$$\tfrac{3}{2}C + \tfrac{3}{2}O_2 = \tfrac{3}{2}CO_2, \qquad \Delta H = -591{\cdot}6\ \text{kJ},$$

$$\tfrac{3}{2}CO_2 + \tfrac{3}{2}C = 3CO, \qquad \Delta H = +258{\cdot}6\ \text{kJ},$$

$$Fe_2O_3 + \tfrac{1}{3}CO = \tfrac{2}{3}Fe_3O_4 + \tfrac{1}{3}CO_2, \qquad \Delta H = -17{\cdot}6\ \text{kJ},$$

$$\tfrac{2}{3}Fe_3O_4 + \tfrac{8}{3}CO = 2Fe + \tfrac{8}{3}CO_2, \qquad \Delta H = -10{\cdot}5\ \text{kJ}.$$

It will be seen that if all these equations are added together (or follow one another consecutively) we have the overall reaction—and according to Hess's law, the heat of the overall reaction will be the sum of the heats of reaction of the stages in the indirect route.

$$\Delta H \text{ for the overall reaction} = +258{\cdot}6 - (591{\cdot}6 + 17{\cdot}6 + 10{\cdot}5)$$

$$= -361{\cdot}1\text{kJ}.$$

This type of assessment gives an indication of the importance of the combustion of carbon to carbon dioxide at the tuyères in supplying sufficient heat to melt the iron, and for other reactions, without there being any need to heat the iron blast furnace externally. (The figures quoted for heats of reaction are for 25°C in this case, but a similar calculation could be made at any temperature, provided the data were available.)

The power of this approach can be appreciated when data are available in the literature as a result of experiment for a large number of reactions, but are not available for a reaction being considered.

To take a second example, the heat of reaction at 25°C for $Cu_2S + 2Cu_2O = 6Cu + SO_2$, a reaction which might take place in a copper converter, could be calculated by the application of Hess's law to the values of the heats of formation of various compounds taken from a table of reference as follows:

Substance	Heat of formation at 25°C[5]
Cu_2S	−82·0 kJ
Cu_2O	−167·4 kJ
Cu	0 (element)
SO_2	−296·9 kJ

The heat of reaction is then

$$0+(-296{\cdot}9)-(-82{\cdot}0)-2(-167{\cdot}4)=+119{\cdot}7\text{ kJ}.$$

Before it is considered that this method is foolproof, the *errors* in the result must be considered. It must be remembered that where heats of reaction are added and subtracted in this way, the experimental error in each result must be added to the rest to give the error in the overall heat of reaction. In the last example quoted, the errors in the heats of formation of Cu_2S, Cu_2O and SO_2 are $\pm 1{\cdot}7$, $\pm 2{\cdot}9$, $\pm 0{\cdot}4$ kJ respectively, so that the error in the heat reaction will be $(1{\cdot}7)+2(2{\cdot}9)+0{\cdot}4$, or $\pm 7{\cdot}9$ kJ. This represents an error of $\pm 6{\cdot}6\%$ in the final result compared with a maximum of $\pm 2\%$ in any of the heats of formation obtained from the tables. It will be seen that large errors can quickly build up in this way, and Hess's law must be applied with caution, bearing in mind the errors in results quoted in the tables especially where the result is a small enthalpy change and has been determined by the difference between large heats of formation.

The heat of reaction alters as the temperature changes, and also the reactants and products may undergo a physical change as the temperature increases—such factors must be taken into account and, to show how this may be carried out, we will use the following example, using data from Kubaschewski and Evans.[5]

In the extraction of zinc by carbon reduction of zinc oxide sinter, the basic reaction could be

$$ZnO_S + C_S = Zn_S + CO_G.$$

At room temperature, the reactants and products would be in the physical states shown in the equation, but the reaction will not proceed at room temperature and the reduction is carried out at about 1100°C. Zinc melts at 420°C and boils at 907°C, so that assuming that the reaction is carried out at 1 atm pressure, the reaction will be

$$ZnO_S + C_S = Zn_G + CO_G,$$

as none of the other reactants changes its physical state in this temperature range. The problem is, given the heats of formation of the reactants and products at 25°C, the relationship between their heat capacities and temperature in the range 25–1100°C, and the latent heats of fusion and evaporation of zinc, calculate the heat of reaction at 25°C and at 1100°C.

Heats of Formation at 25°C:

ZnO_S	$\Delta H_{298} = -348{,}100$ J,
C_S	$\Delta H_{298} = 0$,
Zn_S	$\Delta H_{298} = 0$,
CO_G	$\Delta H_{298} = -110{,}500$ J.

Heat Capacities at Constant Pressure C_p in Joules per Degree C per Mole, with Temperature Range to which they Apply:

ZnO_S, C_p, $ZnO = 48{\cdot}99 + 5{\cdot}10 \times 10^{-3}T - 9{\cdot}12 \times 10^{5}T^{-2}$ (298–1373°K)

C_S, C_p, $C = 17{\cdot}15 + 4{\cdot}27 \times 10^{-3}T - 8{\cdot}79 \times 10^{5}T^{-2}$ (298–1373°K)

Zn_S, C_p, $Zn(S) = 22{\cdot}38 + 10{\cdot}04 \times 10^{-3}T$ (298–m.p. 693°K)

Zn_L, C_p, $Zn(L) = 31{\cdot}38$ (693–b.p. 1180°K)

Zn_G, C_p, Zn(G)=20·79, assuming that zinc is a monatomic gas behaving ideally, so that $C_p = \frac{5}{2}\boldsymbol{R}$, where

$$\boldsymbol{R} = 8{\cdot}314 \text{ J/°K mol} \quad (1180\text{–}1373\text{°K}).$$

CO_G, C_p, $CO = 28{\cdot}41 + 4{\cdot}10 \times 10^{-3}T - 0{\cdot}46 \times 10^5\, T^{-2}$ (298–1373°K).

Heats of Transformation of Zinc:
L_f,Zn =7280 J/mol at 693°K.
L_e,Zn =114,200 J/mol at 1180°K.
To calculate the heat of reaction at 1100°C, we use the integrated form of Kirchhoff's equation (1.18),

$$\Delta H_2 - \Delta H_1, = \int_{T_1}^{T_2} \Delta C_p \,.\, \mathrm{d}T,$$

where ΔC_p is the difference between the heat capacity of the products and the heat capacity of the reactants at constant pressure. Where there is a change in state in one of the products or reactants, the heat of transformation L_t must be taken into account—*subtracted* from the total if a *reactant* transforms, *added* if a *product* transforms. Kirchhoff's equation must then be applied separately above the transformation temperature, including the value for the heat capacity of the substance in its new physical state, so that in general

$$\Delta H_2 - \Delta H_1 = \int_{T_1}^{T_t} \Delta C_{p,\mathrm{I}} \,.\, \mathrm{d}T \pm L_t + \int_{T_t}^{T_2} \Delta C_{p,\mathrm{II}} \,.\, \mathrm{d}T, \quad (1.20)$$

where T_t is the transformation temperature, and $\Delta C_{p,\mathrm{I}}$ and $\Delta C_{p,\mathrm{II}}$ are the values of ΔC_p below and above the transformation temperature respectively.

Here, then, as zinc is a product, the heats of transformation are added and

$$\Delta H_{1373} - \Delta H_{298} = \int_{298}^{693} \Delta C_{p,\mathrm{I}} \,.\, \mathrm{d}T + L_{\mathrm{f}},\mathrm{Zn} + \int_{693}^{1180} \Delta C_{p,\mathrm{II}} \,.\, \mathrm{d}T$$

$$+ L_{\mathrm{e}},\mathrm{Zn} + \int_{1180}^{1373} \Delta C_{p,\mathrm{III}} \,.\, \mathrm{d}T. \qquad (1.21)$$

To calculate ΔH_{298}

$$\Delta H_{298} = (\Delta H_{\mathrm{Zn}} + \Delta H_{\mathrm{CO}}) - (\Delta H_{\mathrm{ZnO}} + \Delta H_{\mathrm{C}}) \qquad \text{(Hess's law)}$$
$$= (0 - 110{,}500) - (-348{,}100 + 0)$$
$$= +237{,}700 \text{ J.}$$

To calculate $\int_{298}^{693} \Delta C_{p,\mathrm{I}} \,.\, \mathrm{d}T$

$$\Delta C_{p,\mathrm{I}} = (C_p,\mathrm{Zn(S)} + C_p,\mathrm{CO}) - (C_p,\mathrm{ZnO} + C_p,\mathrm{C}).\dagger$$
$$= (22{\cdot}38 + 10{\cdot}04 \times 10^{-3}T + 28{\cdot}41 + 4{\cdot}10 \times 10^{-3}T - 0{\cdot}46$$
$$\times 10^5 T^{-2}) - (48{\cdot}99 + 5{\cdot}10 \times 10^{-3}T - 9{\cdot}12 \times 10^5 T^{-2}$$
$$+ 17{\cdot}15 + 4{\cdot}27 \times 10^{-3}T - 8{\cdot}79 \times 10^5 T^{-2})$$
$$= -15{\cdot}36 + 4{\cdot}77 \times 10^{-3}T + 17{\cdot}45 \times 10^5 T^{-2}$$

$$\therefore \int_{298}^{693} \Delta C_{p,\mathrm{I}} \,.\, \mathrm{d}T = \int_{298}^{693} (-15{\cdot}36 + 4{\cdot}77 \times 10^{-3}T + 17{\cdot}45 \times 10^5 T^{-2})\mathrm{d}T.$$

†*Note:* if two molecules of a substance are involved in the equation, the value of C_p for the substance should be doubled.

$$= \left[-15{\cdot}36T + 2{\cdot}38\times10^{-3}T^2 - 17{\cdot}45\times10^5T^{-1}\right]_{298}^{693}$$

$$= [-12{,}000] - [-10{,}200]$$

$$= -1800 \text{ J.}$$

To calculate $\int_{693}^{1180} \Delta C_{p,\text{II}}\,.\,\mathrm{d}T$

$\Delta C_{p,\text{II}}$ is the same as $\Delta C_{p,\text{I}}$ except that C_p,Zn(S) is replaced by C_p,Zn(L).

$$\therefore\ \Delta C_{p,\text{II}} = -6{\cdot}36 - 5{\cdot}27\times10^{-3}T + 17{\cdot}45\times10^5T^{-2}.$$

$$\therefore\ \int_{693}^{1180} \Delta C_{p,\text{II}}\,.\,\mathrm{d}T = \int_{693}^{1180} (-6{\cdot}36 - 5{\cdot}27\times10^{-3}T + 17{\cdot}45\times10^5T^{-2})\mathrm{d}T$$

$$= \left[-6{\cdot}36T - 2{\cdot}64\times10^{-3}T^2 - 17{\cdot}45\times10^5T^{-1}\right]_{693}^{1180}$$

$$= -4460 \text{ J.}$$

To calculate $\int_{1180}^{1373} \Delta C_{p,\text{III}}\,.\,\mathrm{d}T$

$\Delta C_{p,\text{III}}$ is the same as $\Delta C_{p,\text{II}}$ except that C_p,Zn(L) is replaced by C_p,Zn(G).

$$\therefore\ \Delta C_{p,\text{III}} = -16{\cdot}95 - 5{\cdot}27\times10^{-3}T + 17{\cdot}45\times10^5T^{-2}.$$

$$\therefore \int_{1180}^{1373} \Delta C_{p,\mathrm{III}} \, . \, \mathrm{d}T = \int_{1180}^{1373} (-16{\cdot}95 - 5{\cdot}27 \times 10^{-3}T + 17{\cdot}45 \times 10^{5}T^{-2})\mathrm{d}T$$

$$= \Big[-16{\cdot}95T - 2{\cdot}64 \times 10^{-3}T^{2} - 17{\cdot}45 \times 10^{5}T^{-1} \Big]_{1180}^{1373}$$

$$= [-\ 29{,}510] - [-\ 25{,}160]$$

$$= -4350 \text{ J.}$$

Thus, from (1.21),

$$\Delta H_{1373} - \Delta H_{298} = \int_{298}^{693} \Delta C_{p,\mathrm{I}} \, . \, \mathrm{d}T + L_{f},\mathrm{Zn} + \int_{693}^{1180} \Delta C_{p,\mathrm{II}} \, . \, \mathrm{d}T$$

$$+ L_{e},\mathrm{Zn} + \int_{1180}^{1373} \Delta C_{p,\mathrm{III}} \, . \, \mathrm{d}T,$$

and therefore

$$\Delta H_{1373} - (+237{,}700) = (-1800) + (+7280) + (-4460) + (+114{,}200) + (-4350)$$

and

$$\Delta H_{1373} = 237{,}700 + 110{,}900$$

$$= +348{,}600 \text{ J.}$$

Thus, at 1100°C, the reaction is even more endothermic than at 25°C, and the major factor in raising the heat of reaction to this value is the latent heat of evaporation of zinc, +114,200 J/mol.

The commercial importance of the knowledge that this reaction is so strongly endothermic is that heat supply and conservation in the reduction of zinc oxide sinter by carbon—whether in horizontal retorts, vertical retorts or the blast furnace—are of great importance, and it was this factor combined with the more important problem of the difficulty of condensation of the zinc vapour as it left the smelting unit which delayed the successful commercial introduction of the comparatively large unit of the blast furnace until 1950.

It is hoped that the presentation of this calculation in full detail will illustrate the foregoing discussion of the theory of enthalpy changes and their relationship to heat capacities, and will give some idea of the type of calculation involved when some thermodynamic problems are tackled. If the range of temperature over which the change in ΔH is to be determined is small, then the variation in heat capacities with temperature can be ignored, and the calculation becomes more simple. Tables of values of $(H_T - H_{298})$ for elements and compounds —H_T and H_{298} being the heat contents at temperature T°K and 298°K respectively—have been compiled to reduce the amount of calculation necessary where the temperature range is sufficiently large to alter C_p significantly.[6] Where accuracy of experimental data is insufficient to warrant the use of more than one temperature-correction term, expressions such as (1.15) can be shortened to more convenient dimensions, such as $C_p = \mathrm{A} + \mathrm{B}T$.

1.10. Experimental Techniques in Calorimetry

It is not intended that this book will include a detailed review of experimental techniques employed in *calorimetry*—the measurement of quantities of heat energy—and for such a review, with special reference to metallurgical systems, the reader should consult ref. 5, Part II, Section 1, and ref. 7, Chapter 7. Here, a brief survey, indicating the special metallurgical problems encountered will be given.

The first essential in calorimetry is an instrument which will measure temperature sufficiently accurately and with maximum convenience. The three main techniques used have been mercury-in-glass thermometers, platinum resistance thermometers and thermocouples. The mercury-in-glass thermometer must have a fine bore and a large mercury reservoir to give the necessary accuracy—either a set of thermometers with overlapping ranges each measuring a range of about 5°C, or a Beckmann thermometer, which has a range of 6°C, but whose zero can be adjusted between −10°C and 100°C, can be used. Such instruments can be read to 0·001°C and are relatively cheap and simple to operate. Platinum resistance thermometers give over 5 times the accuracy of the mercury-in-glass thermometers, depending on the electrical measuring instruments employed, and the platinum resistance wire can be built into the structure of the calorimeter, care being taken to eliminate strains and mount the element with a good insulator. Thermocouples are usually used in series with several other thermocouples to give increased sensitivity—and this also allows a number of couples to be distributed over large bodies to read an average temperature of the body. Problems of increased error due to the thermal conductance of the connections, and thermal lag of the couple sheaths are encountered. Radiation pyrometers can be used where it is difficult to approach the object being heated, a rapid response is required and there is to be no heat conduction from the system via the instrument, but in general their accuracy and sensitivity are less than those of the other three techniques. A chapter on temperature measurement, with a comprehensive list of references, is included in ref. 7, Chapter 2.

The next essential is knowledge of the heat capacity of the calorimeter—this includes the heat capacities of the inner containing vessel, temperature measuring instruments, and other surroundings contributing in any measurable way to the heat changes in the system under observation. All these are included in *calorimeter constant* which is the amount of heat required to raise the temperature of the calorimeter by 1 degree (often known as "water equivalent" of the calorimeter), which is measured by supplying a known amount of heat to the calorimeter, either by electrical heating or dropping a piece of metal of known heat capacity into the calorimeter and measuring the resulting temperature change. Such experiments must be carried out in the temperature range and at about the same rate of heat evolution to be used in the subsequent calorimetric determinations.

The three quantities which are to be measured are indicated by the integrated Kirchhoff equation (1.20)—heats of reaction (combustion, solution, etc.), heat capacities, and heats of transformation. If necessary, the heat of reaction can be measured directly at the temperature required, or alternatively at room temperature so that the heat of reaction at any other temperature can be calculated knowing the necessary values of heat capacities and heats of transformation as in the example quoted in Section 1.9. Indirect measurements of heats of reaction can be carried out by using simpler reactions and computing the required heat of reaction by applying Hess's law (Section 1.9). It is usually more convenient to carry out measurements of heats of reaction at room temperature, but where a reaction is very slow it may be necessary to carry it out at a high temperature to increase the speed, for example, where solids are reacting.

Two distinct methods can be employed—isothermal and adiabatic—the former requiring a constant temperature jacket and the latter a jacket whose temperature can be adjusted so that it is the same as the contents of the vessel, thus preventing any heat exchange between the system and its surroundings. In isothermal calorimetry, corrections must be made for the exchange of heat between the calorimeter and its surroundings during the reaction, whereas this is inapplicable to adiabatic calorimetry, making it more suitable perhaps for high temperature calorimetry where errors due to heat losses are increased. The drawbacks of adiabatic calorimetry are the difficulties associated with compensating for sudden changes in temperature in the calorimeter.

The "bomb calorimeter" is a useful piece of apparatus where the reaction involves a gas—particularly for determination of heats of combustion. The bomb is a strong gas-tight steel vessel in which the material to be oxidized, which may be a liquid or a solid, is placed in a small crucible supported by thin wires, which may be used to carry the electric current which ignites the material. Oxygen is pumped into the bomb at up to 25 atm pressure, and the bomb is immersed in a calorimeter full of water—all enclosed in a constant temperature jacket. The change in temperature of the calorimeter after ignition of the bomb is measured, and the apparatus is calibrated using a substance of known heat of combustion or by an electrical method. Heats of reaction involving chlorine and nitrogen have also been determined by this method.

Heat capacities can be determined by measuring the change in heat content of a substance when it is heated to an elevated temperature and dropped into a calorimeter in a constant temperature vessel. This is the "method of mixtures" commonly used in teaching the determination of specific heats (see Fig. 1.6). Care has to be taken to prevent oxidation or decomposition of the material while it is being heated, in the

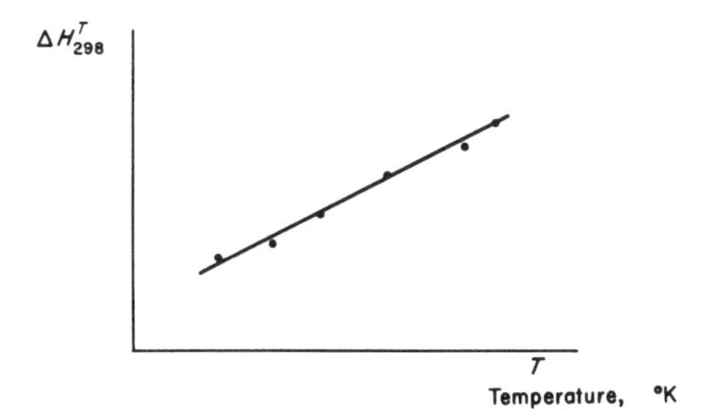

FIG. 1.6. "Method of mixtures" for determination of heat capacities. ΔH^T_{298} is the change in heat content (enthalpy) of a substance when cooled from T°K to 298°K. The slope of the graph gives C_p, the heat capacity at constant pressure, over the temperature range of the graph.

case of reactive metals for example, and also the material in the calorimeter must not evaporate, or react with the substance being dropped into it. Direct measurement of heat capacities is achieved by measuring the rise in temperature of a specimen when it is heated by a known amount of electrical energy. Enough heat is supplied to raise the temperature by a few degrees, and this technique can be used at any temperature—either isothermally or adiabatically.

Heats of transformation can be measured in the same way as heat capacities, by the method of mixtures, from a curve such as Fig. 1.7. This method is more suitable for heats of

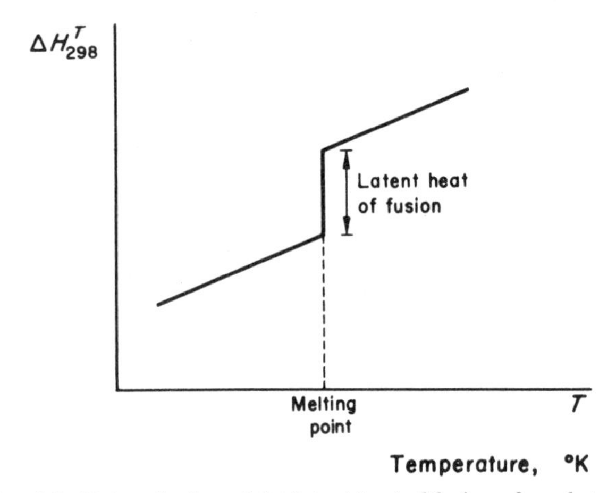

FIG. 1.7. Determination of the latent heat of fusion of a substance by the "method of mixtures". ΔH_{298}^{T} is the change in heat content (enthalpy) of the substance when it is cooled from T°K to 298°K.

fusion than for heats of transformation in the solid state because the rapid rate of cooling might inhibit the transformation, particularly in diffusion-controlled transformations. Direct measurements of heats of transformation can be carried out by a similar technique to those for heat capacities, and the heats of order–disorder transformations and precipitation reactions in solid alloys have been determined.

Where high temperature calorimetric measurements are to be made, problems of uniform temperature control and refractory materials for construction of vessels to hold reactive materials at high temperatures become very important. Any reaction with the container would introduce errors in measurements of heats of reaction and heat capacities, and the temperature at which the reaction is carried out must be maintained throughout a reasonable volume of the calorimeter. Other problems include increased heat losses by radiation and a decrease in the thermal and electrical resistance of ceramic materials.

References

1. GLASSTONE, S. *Textbook of Physical Chemistry*, Macmillan, London, 1953.
2. BUTLER, J. A. V. *Chemical Thermodynamics*, Macmillan, London, 1951: LEWIS, G. N. and RANDALL, M. (revised by K. S. Pitzer and L. Brewer) *Thermodynamics*, McGraw-Hill, London, 1961: OBERT, E. F. and GAGGIOLI, R. A. *Thermodynamics*, McGraw-Hill, New York, 1963.
3. A discussion of the theories of Debye (1912) and others can be found in ref. 1 and ref. 4.
4. DARKEN, L. S. and GURRY, R. W. *Physical Chemistry of Metals*, McGraw-Hill, New York, 1953.
5. KUBASCHEWSKI, O. and EVANS, E. LL. *Metallurgical Thermochemistry*, Pergamon, London, 1958.
6. *Basic Open Hearth Steelmaking*, A.I.M.E., 3rd edn., 1964, p. 543.
7. BOCKRIS, J. O'M., WHITE, J. L. and MACKENZIE, J. D. *Physico-Chemical Measurements at High Temperatures*, Butterworths, London, 1959.

CHAPTER 2

Entropy, Free Energy and Chemical Equilibrium

2.1. Introduction

In the first chapter, we were mainly concerned with an experimental law—the First Law of Thermodynamics—and its implications. In this chapter we shall again be considering the results of experiment, and these can be introduced by the following facts, which are both statements of the Second Law of Thermodynamics:

(i) heat always flows from a hotter to a colder body—never in the reverse direction;

(ii) an isolated system always tends to take up a more *disordered* form—never of its own accord becoming more ordered.

We would be surprised to find that if two ingots—one at 100°C and the other at 500°C—were placed close to one another in a soaking pit with no other source of heat, the hotter ingot increased in temperature until it melted, whereas the colder ingot cooled down to 0°C. Instead, the temperature of the hotter ingot would *always* decrease as heat flowed out from it (whether by radiation, convection or conduction) to raise the temperature of the colder ingot.

If we consider a pattern formed by coloured counters on a flat board (Fig. 2.1a), we know that if the counters were picked up and thrown down again, they would form a pattern of the type shown in Fig. 2.1b. We say that the counters in (a) are *more ordered* than those in (b). If the counters in (b) were picked up and thrown down again it would be very improbable that they would fall in the pattern (a). We know that state (b) is *more probable* than state (a), and can calculate, by the techniques of statistics, the "probability" W of the two systems. We know that $W_{(b)} > W_{(a)}$, and it is an experimental fact that, in any change which is dependent only on the laws of chance, the change will be such that the *probability* of the state of the system will increase. A more disordered system is more probable than a more ordered system, so that changes of this type are accompanied by an increase in disorder of the system.

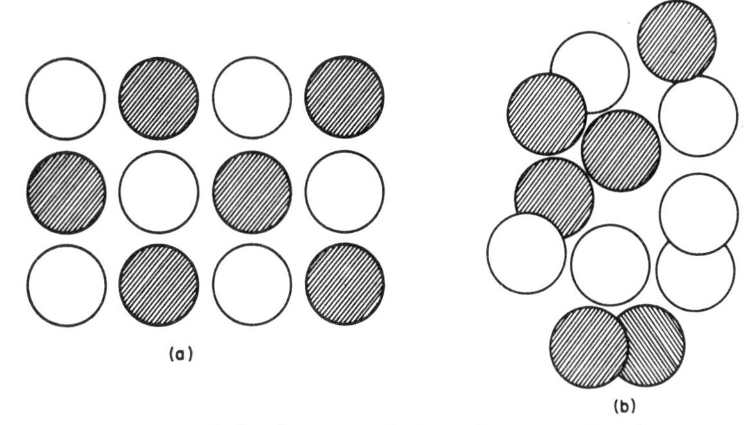

FIG. 2.1. (a) Ordered pattern of coloured counters placed on a flat board. (b) Disordered pattern of coloured counters formed if the counters in (a) were picked up and thrown down again on the board.

A physical example of this experience is shown by the molecules of a gas enclosed in a chamber which is connected by a tap to an evacuated chamber (Fig. 2.2a). If the tap is opened, the gas molecules, by their random motion in the space available to them, will enter the second chamber until, on the average, there will be the same number of molecules in each chamber (Fig. 2.2b). In other words, a gas always tends to "fill" any chamber in which it is placed. We know from

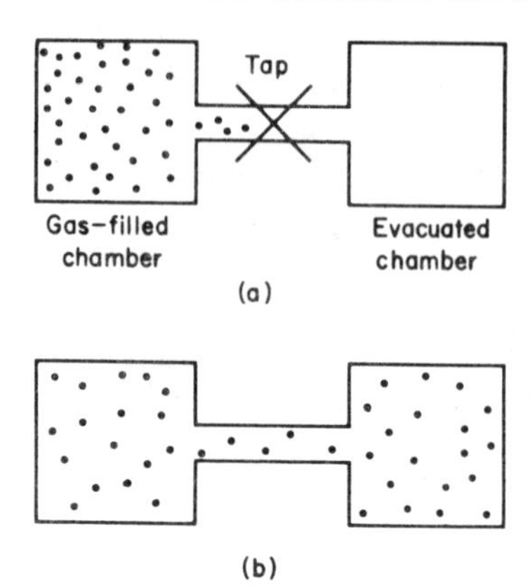

FIG. 2.2. The movement of gas molecules into an evacuated chamber.

experience that the molecules in the right-hand chamber will never of their own accord all move into the left-hand chamber leaving the right-hand chamber empty. Condition (b) is *more probable* than condition (a), and is *more disordered* than condition (a). A "spontaneous change" has taken place, accompanied, as always, by an increase in probability – and therefore an increase in the disorder of the system.

2.2. The Statistical Nature of the Second Law of Thermodynamics

The phrase "on the average", used in Section 2.1, is important. We know that if we were able to watch one of the molecules in the gas system, we might see it pass from one chamber to the other, and back, through the connecting tube – moving against the "spontaneous change" which was taking place in the system. Only if we consider all the molecules in the system – by means of a statistical average – is this Second Law obeyed. Thermodynamics is not concerned with the behaviour of the individual molecules of a gas system – only the *statistical average behaviour* of the system. The pressure exerted by a gas is the resultant of the distribution of forces produced by the motion and collision of many molecules—each with its own individual energy—and in considering the pressure of a gas, we do not watch the effect of each individual molecule—only that of the motion of all the molecules.

In addition to an energy change, we can therefore have a change in the state of order accompanying a chemical or physical change in a system, and this change in the state of order can be expressed as a change in a property of the system, **the entropy** S.

L. Boltzmann (1877) proposed that

$$S = k \ln W + S_0 \tag{2.1}$$

where S is the *entropy* of the system, $\boldsymbol{k}$ is Boltzmann's constant (the gas constant $\boldsymbol{R}$, divided by Avogadro's Number $\boldsymbol{N}$, is $\boldsymbol{k}$, the gas constant per single molecule), W is the *thermodynamic probability* of the system, and S_0 is a constant—the entropy when W is unity.

The *thermodynamic probability* of the system, W, is defined as the ratio of the probability of the existing state of the system to the probability of the state of complete order in the system for the same energy and volume.

The Third Law of Thermodynamics is concerned with the value of S_0—because it has been argued by many workers (Nernst, Planck, Simon) that a perfect crystalline solid should be completely ordered at the absolute zero of temperature—and should therefore have zero entropy. The Third Law assumes this, giving S_0 the value of zero because in a condition of perfect order, $W = 1$, and $S = S_0 = 0$.

Thus we have

$$S = \boldsymbol{k} \ln W. \tag{2.2}$$

From this relationship we see that, as the system becomes more disordered, W increases and S increases. The entropy of the system can be used as a measure of the state of order, or probability, of the system. For a full development of this relationship between entropy and probability, the reader should consult ref. 1 or 2, or a textbook on statistical thermodynamics.

We could now state the Second Law of Thermodynamics as follows: a spontaneous process is always accompanied by an increase in the total entropy of the system and its surroundings. This in fact says virtually the same thing as statements (i) and (ii) of the Law in Section 2.1.

Considering the two ingots—one hot and one cold—placed in a soaking pit, the particles which make up the hot ingot will be relatively disordered, and those of the cold ingot relatively ordered. The hot ingot will have higher entropy than the cold ingot. When heat flows from the hot ingot to the cold ingot, the entropy of the cold ingot will increase, and that of the hot ingot will decrease—but in such a way that the total entropy of the system increases. This continues until both ingots are at the same temperature and we reach equilibrium—the state of maximum entropy. (It must be remembered that the entropy of a *system and its surroundings* will increase—and here we have neglected, wrongly, the entropy increase of the materials of construction of the soaking pit when heat flows from the hot ingot to the colder parts of the soaking pit.) This phenomenon will be demonstrated quantitatively at the end of Section 2.4.

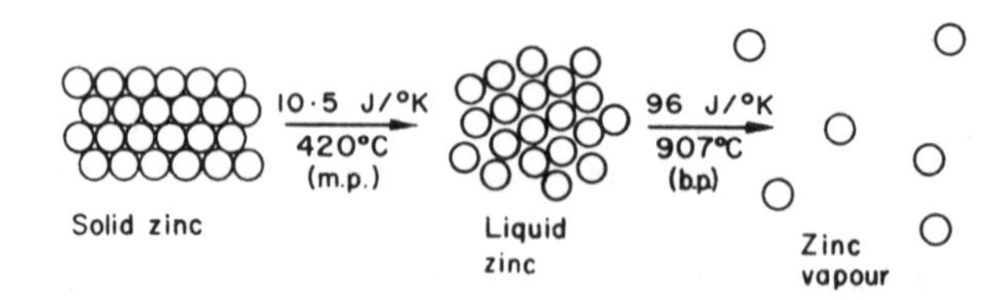

FIG. 2.3. Representation of the changes in the structure of zinc when it melts and boils. Solid zinc has a close-packed hexagonal crystal structure, and an entropy increase of 10·5 J/°K accompanies the fusion to form a slightly broken crystal lattice in the liquid state. On boiling, the atoms become completely separated with an increase in entropy of 96 J/°K.

To take one more example, the melting and boiling of zinc (Fig. 2.3); at its melting point, the close-packed hexagonal crystal structure of zinc becomes slightly broken up to produce a liquid – with an entropy increase (ΔS) of 10·5 J/°K. At its boiling point, the liquid structure is completely destroyed to form the completely random gaseous phase, and the entropy increases by 96 J/°K. We know that the change in the state of order of the system at the boiling point is much greater than at the melting point (the volume of a gas is much greater than that of a liquid) – and this is reflected in the very much larger entropy change on boiling. (We will discuss the units used for entropy in the next section.)

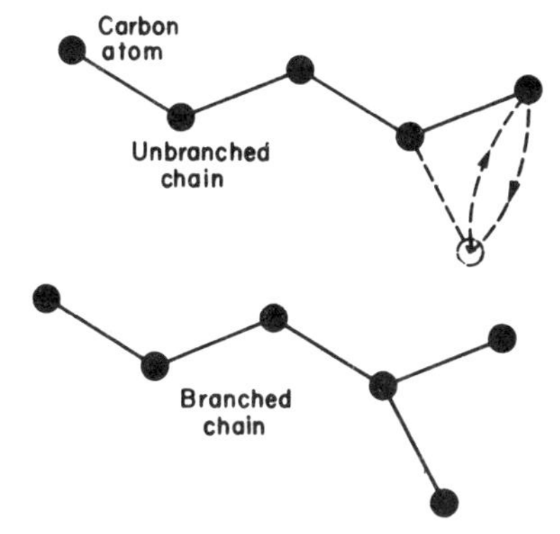

FIG. 2.4. Rotation of one carbon atom with respect to another in an unbranched carbon chain. When the chain becomes branched, the rotation of the carbon atom in a branch becomes restricted by the presence of the carbon atom in the other branch.

In addition to the disorder produced by motion of the molecules in a system (entropy of translation), we also have the rotation and vibration within the structure of each molecule. The more ways in which a molecule can dissipate its energy, the greater its entropy. Thus a long flexible chain molecule (like a hydrocarbon $CH_3{\cdot}CH_2{\cdot}CH_2{\cdot}CH_2{\cdot}CH_2{\cdot}CH_2{\cdot}CH_2{\cdot}CH_2{\cdot}CH_3$) will have many different positions and rotations of the carbon atoms with respect to one another (Fig. 2.4), whereas a branched isomer of the same formula (C_9H_{20}) will be more restricted in the way in which its carbon atoms can rotate about the axis of the molecule,

$$CH_3{\cdot}CH_2{\cdot}CH_2{\cdot}CH_2{\cdot}CH_2{\cdot}CH_2{\cdot}CH\begin{matrix}\diagup CH_3\\ \diagdown CH_3\end{matrix},$$

because of the hindrance in space caused by the branching of the chain. In fact, as a hydrocarbon chain becomes more branched, the entropy of that substance decreases because of the reduction in the randomness of positions of atoms within the molecule.

2.3. A Different Approach to the Entropy Function: Cyclic Processes

Entropy measurements, if they could be made, would apparently give us some information about the structure of molecules, of solids and of liquids—but how can this be done? What are the "entropy units" mentioned in the last section?

"Classical" thermodynamics provides the answer—in fact, historically, this approach came before that of statistical thermodynamics. The existence of an entropy function was postulated by the thermodynamicists before its behaviour and significance were expanded by the statistical mathematicians.

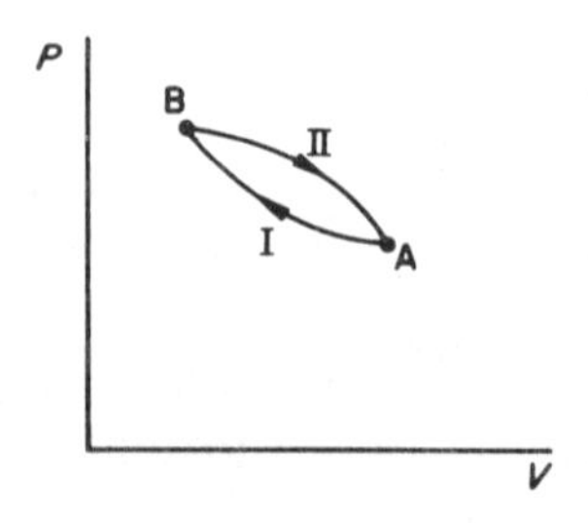

FIG. 2.5. A cycle: the system, initially at A, is taken to B via path I, then returns to A via path II. P is the pressure of the system, V the volume.

Figure 2.5 shows a "cycle", in which the system is changed from A to B and back again—even though by two different routes I and II. We know from the First Law of Thermodynamics and the nature of thermodynamic variables (Section 1.4) that the change in internal energy ΔU will be zero and that values of P and V will not change as a result of a system being taken through such a cycle. Therefore the work done by the system, w, must be the same as the heat absorbed by the system, q, in traversing such a cycle.

$$\Delta U = q - w, \qquad \text{from (1.2).}$$

$$\therefore q = w, \text{ as } \Delta U = 0.$$

If we have some sort of machine, or "heat engine", which can convert heat into mechanical work (e.g. a steam engine), we can define the *efficiency* of the machine as the fraction of heat absorbed by the machine which can be converted into work.

S. Carnot (1824) proposed a theorem which stated that all machines working thermodynamically reversibly in cycles between the same temperatures of source and sink have the same efficiency. By thermodynamically reversible working, Carnot's theorem refers to changes of the type discussed in Section 1.5, and the "temperatures of source and sink" refer, in the case of a steam engine, to the temperature of the steam supplied to the engine and the temperature of the steam when the engine has completed its work respectively.

Carnot's cycle is a useful method of calculating the efficiency of heat engines. Since, according to this theorem, all reversible machines have the same efficiency, we need consider only one simple machine, in which the working substance is 1 mole of a perfect gas contained in a cylinder with a frictionless, weightless piston of the type discussed in Section 1.4. The cycle is shown in Fig. 2.6, where A is the starting point, and it is carried out as follows:

Stage I. The gas, at temperature T_1, pressure P_A and volume V_A is allowed to expand isothermally and reversibly to B, where its pressure is P_B, volume V_B.

From (1.10), the work done by the gas is

$$w_I = RT_1 \ln \frac{V_B}{V_A} = q_I, \text{ the heat absorbed by the gas.}$$

Stage II. The gas is then expanded adiabatically (with no heat exchange with its surroundings) to C, volume V_C, pressure P_C. There will be a fall in temperature to T_2 because the work done is at the expense of a fall in kinetic energy of the gas molecules. The energy produced when a gas cools under these conditions is related to the molar heat capacity at constant volume C_v as follows (see Section 1.8):

$$w_{II} = C_v(T_1 - T_2).$$

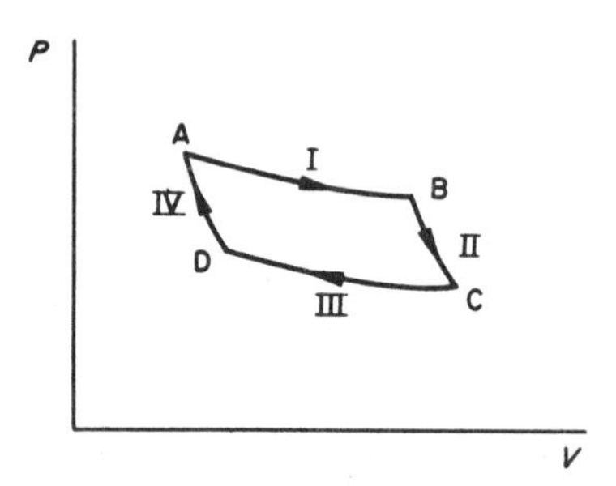

FIG. 2.6. Carnot's cycle.

Stage III. The gas is compressed isothermally to D. Then the work done by the gas,

$$w_{III} = RT_2 \ln \frac{V_D}{V_C} = q_{III}.$$

Stage IV. To complete the cycle, the gas is compressed adiabatically to A, and the work done by the gas is

$$w_{IV} = C_v(T_2 - T_1).$$

Therefore the total work done by the gas in completing the cycle is

$$w = w_{\text{I}} + w_{\text{II}} + w_{\text{III}} + w_{\text{IV}}.$$

Thus,

$$w = \boldsymbol{R}T_1 \ln \frac{V_{\text{B}}}{V_{\text{A}}} + C_v(T_1 - T_2) + \boldsymbol{R}T_2 \ln \frac{V_{\text{D}}}{V_{\text{C}}} + C_v(T_2 - T_1),$$

$$= \boldsymbol{R}T_1 \ln \frac{V_{\text{B}}}{V_{\text{A}}} + \boldsymbol{R}T_2 \ln \frac{V_{\text{D}}}{V_{\text{C}}}$$

$$= q_{\text{I}} + q_{\text{III}}, \quad \text{the net heat absorbed by the system.}$$

For an adiabatic expansion, between temperatures T_1 and T_2, it is known that, if V_1 is the volume at T_1, and V_2 the volume at T_2, from (1.19)

$$\frac{T_1}{T_2} = \left(\frac{V_2}{V_1}\right)^{(\gamma-1)}$$

where $\gamma = C_p/C_v$, the ratio of the molar heat capacities at constant pressure and constant volume (see Section 1.8).

But stages II and IV are adiabatic expansions between the same temperatures T_1 and T_2, so that

$$\frac{T_1}{T_2} = \left(\frac{V_{\text{D}}}{V_{\text{A}}}\right)^{(\gamma-1)} = \left(\frac{V_{\text{C}}}{V_{\text{B}}}\right)^{(\gamma-1)}.$$

Therefore $V_{\text{D}}/V_{\text{A}} = V_{\text{C}}/V_{\text{B}}$, and $V_{\text{D}}/V_{\text{C}} = V_{\text{A}}/V_{\text{B}}$,

and

$$w = \boldsymbol{R}T_1 \ln \frac{V_{\text{B}}}{V_{\text{A}}} + \boldsymbol{R}T_2 \ln \frac{V_{\text{A}}}{V_{\text{B}}}$$

$$= \boldsymbol{R}(T_1 - T_2) \ln \frac{V_{\text{B}}}{V_{\text{A}}}.$$

Hence,

$$\text{Efficiency of the heat engine} = \frac{w}{q_{\mathrm{I}}} = \frac{\boldsymbol{R}(T_1 - T_2)\ln(V_{\mathrm{B}}/V_{\mathrm{A}})}{\boldsymbol{R}T_1 \ln(V_{\mathrm{B}}/V_{\mathrm{A}})}$$

$$= \frac{T_1 - T_2}{T_1}. \qquad (2.3)$$

Thus the efficiency of all reversible heat engines is only related to the temperatures of source and sink (T_1 and T_2 respectively), and is independent of the nature of the working substance; this is Carnot's theorem. Note here that no heat engine can be 100% efficient—as, for this to occur, T_2 would have to be the absolute zero of temperature—a practical impossibility.

This approach was the basis of the *thermodynamic scale of temperature* proposed by Lord Kelvin, where the absolute zero of temperature is that value of T_2 which would give an efficiency of 1. The Kelvin scale and the Gas scale are the same, provided that the gas is an ideal gas. The absolute zero of temperature is taken as −273·16°C (degrees Celsius) and absolute temperatures are often referred to as °K (degrees Kelvin).

From the examination of Carnot's cycle, we see that, (2.3)

$$\frac{w}{q_{\mathrm{I}}} = \frac{T_1 - T_2}{T_1}$$

and

$$w = q_{\mathrm{I}} + q_{\mathrm{III}}.$$

Therefore

$$\frac{q_{\mathrm{I}} + q_{\mathrm{III}}}{q_{\mathrm{I}}} = \frac{T_1 - T_2}{T_1},$$

$$1 + \frac{q_{\mathrm{III}}}{q_{\mathrm{I}}} = 1 - \frac{T_2}{T_1},$$

$$\frac{q_{\mathrm{I}}}{T_1} = -\frac{q_{\mathrm{III}}}{T_2},$$

and

$$\frac{q_{\mathrm{I}}}{T_1} + \frac{q_{\mathrm{III}}}{T_2} = 0. \tag{2.4}$$

We can say, in general, that

$$\sum \frac{q}{T} = 0 \tag{2.5}$$

for any reversible cycle, which can be considered as being made up of a large number of Carnot cycles.

Clausius (1865) called this ratio of the heat absorbed by a system, q, to the temperature at which it is absorbed, T, the *entropy change* of the system, where an infinitesimal change in entropy $\mathrm{d}S$ is defined by the relationship

$$\mathrm{d}S = \frac{\delta q}{T}, \tag{2.6}$$

δq being the small amount of heat absorbed by the system and T the absolute temperature at which it was absorbed.

Entropy was later found to bear a relationship to the probability of a system, as discussed in Section 2.2, and expressed in (2.2)

$$S = k \ln W.$$

It is an extensive property of the system as it depends on the mass of the system, and is a *thermodynamic variable*, depending only on the state of the system, not on its history (see Section 1.4). Its dimensions are those of energy divided by temperature and the usual unit is joules per degree Kelvin per mol. As the system absorbs heat energy, its entropy will increase, and we saw that this was the case with the melting and boiling of zinc (Section 2.2).

If the heat engine worked *irreversibly*, we would get less work out of it than if it worked reversibly because reversible changes produce maximum work (Section 1.7), and we can now restate the Second Law of Thermodynamics in the light of the discussion in this section:

No practical heat engine can have 100% efficiency, and heat energy cannot be completely transformed into mechanical energy.

This is the same as the other statements of the Law which have been made, but is rather restricted in its scope compared with the more general statement in Section 2.2, which refers to the increase in entropy accompanying spontaneous changes.

2.4. Some Thermodynamic Relationships Involving Entropy

If we consider the system shown in Fig. 2.5, the entropy of the system at A can be S_A and at B, S_B, so that if the system changes reversibly from state A to state B,

$$\Delta S = S_B - S_A,$$

and is independent of the path taken. If the change were *irreversible*, ΔS would still be the same because S is a thermodynamic variable. In completing the cycle A → B → A,

$$\Delta S = S_B - S_A + S_A - S_B = 0,$$

because we finish up where we started, and there is no net change in entropy. This is to be expected from the relationship (2.5). If the change is *adiabatic*, no heat is absorbed or evolved, ΔS is zero and the change is *isentropic*.

To calculate the change in entropy accompanying a finite change in the system, we must integrate (2.6) between the limits of temperature involved in the change, because ΔS is the sum of all the entropy increments (2.5). For ΔS to be calculated, each increment of heat absorbed δq must be a reversible heat change, and

$$\Delta S = S_2 - S_1 = \int_{T_1}^{T_2} \frac{\delta q}{T}, \tag{2.7}$$

where S_2 and S_1 are the entropies of the system at temperatures T_2 and T_1 respectively.

Now, from (1.12), at constant pressure,

$$C_p = \frac{\delta q_p}{\mathrm{d}T}.$$

Therefore $\mathrm{d}S = \dfrac{\delta q_p}{T} = \dfrac{C_p \mathrm{d}T}{T}$ [from (2.6)]

using (2.7)

$$\Delta S = S_2 - S_1 = \int_{T_1}^{T_2} \frac{\delta q_p}{T} = \int_{T_1}^{T_2} \frac{C_p \mathrm{d}T}{T}. \tag{2.8}$$

Similarly, at constant volume,

$$\Delta S = \int_{T_1}^{T_2} \frac{C_v \mathrm{d}T}{T}. \tag{2.9}$$

If the pressure and volume are not constant, using (1.6) (1.14), for a small heat change,

$$\delta q = \mathrm{d}U + P\mathrm{d}V$$

$$= C_v \mathrm{d}T + P\mathrm{d}V.$$

Therefore,

$$\frac{\delta q}{T} = C_v \frac{\mathrm{d}T}{T} + \frac{P\mathrm{d}V}{T} = C_v \frac{\mathrm{d}T}{T} + \frac{R\mathrm{d}V}{V},$$

because $PV = \boldsymbol{R}T$ for an ideal gas.
Therefore,

$$\Delta S = S_2 - S_1 = \int_{T_1}^{T_2} C_v \frac{\mathrm{d}T}{T} + \int_{V_1}^{V_2} \frac{R\mathrm{d}V}{V}. \qquad (2.10)$$

For an isothermal process, T is constant and

$$\Delta S = \int_{V_1}^{V_2} \frac{\boldsymbol{R}\mathrm{d}V}{V} = \boldsymbol{R} \ln \frac{V_2}{V_1}. \qquad (2.11)$$

We can now calculate the entropy change when the gas in Fig. 2.2a fills the evacuated chamber isothermally. The volume of the gas is doubled so that $V_2/V_1 = 2$ (assuming that the two chambers have equal volume). Using (2.11),

$$\Delta S = \boldsymbol{R} \ln 2 = 8.314 \times 0{\cdot}6931$$
$$= 5{\cdot}762 .$$

Since the gas constant $\boldsymbol{R}$ = 8·314 J/°K mol, the entropy change is 5·762 J/°K mol.

If a phase transformation occurs in the temperature range being considered, there will be an entropy change accompanying that transformation as we saw for zinc in Section 2.2. This will be equal to the heat of transformation (L_t) divided by the transformation temperature T_t (see Section 1.9).

$$\Delta S_t = \frac{L_t}{T_t}. \qquad (2.12)$$

Thus, at constant pressure, from (2.8),

$$\Delta S = S_2 - S_1 = \int_{T_1}^{T_t} \frac{C_p}{T} \mathrm{d}T + \frac{L_t}{T_t} + \int_{T_t}^{T_2} \frac{C_p}{T} \mathrm{d}T, \tag{2.13}$$

where a phase transformation is involved.

Over short temperature ranges, C_p can be assumed to remain constant, but over larger temperature ranges, the relationship between C_p and T, of the type given in (1.15), must be considered:

$$C_p = \mathrm{a} + \mathrm{b}T + \mathrm{c}T^{-2}.$$

For convenience, *standard entropies* of substances are tabulated—that is the entropy at 298°K (25°C)—in tables such as those in ref. 3.

Here $\Delta S = S_{298} - S_0$, and from the Third Law of Thermodynamics, we assume that S_0, the entropy of a substance at 0°K, is zero.

Then we can use these values to calculate the entropy of a substance at temperatures other than 25°C.

Example. To calculate the entropy of zinc at 727°C.

Using the tables in ref. 3, for zinc,

$$S_{298} = 41{\cdot}63 \text{ J/°K mol.}$$

$$L_f = 7280 \text{ J/mol at 420°C.}$$

Solid zinc, $C_{p(\mathrm{S})} = 22{\cdot}38 + 10{\cdot}04 \times 10^{-3}T$ J/mol.
Liquid zinc, $C_{p(\mathrm{L})} = 31{\cdot}38$ J/mol.
Using (2.13),

$$\text{Therefore } \Delta S = S_{1000} - S_{298} = \int_{298}^{693} \frac{C_{p(\mathrm{S})}}{T} \mathrm{d}T + \frac{L_f}{T_f} + \int_{693}^{1000} \frac{C_{p(\mathrm{L})}}{T} \mathrm{d}T$$

$$=\int_{298}^{693}\left(\frac{22\cdot38}{T}+10\cdot04\times10^{-3}\right)\mathrm{d}T+\frac{7280}{693}+\int_{693}^{1000}\frac{31\cdot38}{T}\mathrm{d}T.$$

$$=\Big[22\cdot38\ \ln T+10\cdot04\times10^{-3}T\Big]_{298}^{693}+10\cdot50+\Big[31\cdot38\ \ln T\Big]_{693}^{1000}$$

$$=22\cdot89+10\cdot50+11\cdot46=44\cdot85.$$

Thus, $S_{1000}=44\cdot85+41\cdot63=86\cdot48$ J/°K mol.

Entropy is an extensive property of the system, so that entropy values are additive in the same way as heat contents. The entropy change for a reaction is obtained by a summation of the entropies of reactants and products—for example, in the reaction

$$FeO+C=Fe+CO,$$

$$\Delta S=S_{Fe}+S_{CO}-S_{FeO}-S_{C}.$$

This is similar to a "Hess's law" addition of heat contents such as was discussed in Section 1.9.

When a spontaneous change takes place in a system, the Second Law of Thermodynamics states that the total entropy change of the system and its surroundings is positive. The example of two ingots mentioned in Section 2.2 provides an opportunity to demonstrate that this leads to the experience that heat will flow from a hotter to a colder body but never in the reverse direction. Assuming the transfer of heat from the ingots to the materials of construction of the soaking pit is negligible, any heat transfer considered must be exclusively between the two ingots. Let the ingots be of the same weight, of iron whose molar heat capacity is given by $C_p=17\cdot49+24\cdot77\times10^{-3}T$ J/°K mol,[3] and at a temperature of 400 and

200°C respectively. If heat is transferred from the hot to the cold ingot until the temperature of the two ingots is the same (300°C), the entropy change accompanying this process can be calculated using (2.8),

$$\Delta S = \int_{T_1}^{T_2} \frac{C_p}{T} \, . \, \mathrm{d}T = \int_{T_1}^{T_2} \left(\frac{17{\cdot}49}{T} + 24{\cdot}77 \times 10^{-3}\right)$$

For the hot ingot,

$$\Delta S_{\mathrm{H}} = \int_{673}^{573} \left(\frac{17{\cdot}49}{T} + 24{\cdot}77 \times 10^{-3}\right) . \, \mathrm{d}T = -5{\cdot}289 \text{ J/°K mol.}$$

For the cold ingot,

$$\Delta S_{\mathrm{C}} = \int_{473}^{573} \left(\frac{17{\cdot}49}{T} + 24{\cdot}77 \times 10^{-3}\right) . \, \mathrm{d}T = +5{\cdot}832 \text{ J/°K mol.}$$

Therefore the total entropy change of the system and its surroundings is given by

$$\Delta S_{\mathrm{H}} + \Delta S_{\mathrm{C}} = +0{\cdot}544 \text{ J/°K mol.}$$

This is positive, so that the process is spontaneous according to the Second Law of Thermodynamics. If the hot ingot were to gain heat from the cold ingot so that its temperature rose from 400 to 600°C, while the temperature of the cold ingot dropped from 200 to 0°C,

$$\Delta S_{\mathrm{H}} + \Delta S_{\mathrm{C}} = +9{\cdot}48 - 14{\cdot}52 = -5{\cdot}04 \text{ J/°K mol.}$$

This is negative and the process is not spontaneous—it is a matter of experience that heat will not flow from a colder to a hotter body.

2.5. The Experimental Determination of Entropies

Equation (2.8) can be rewritten as

$$\Delta S = S_2 - S_1 = \int_{T_1}^{T_2} C_p \,.\, \mathrm{d} \ln T,$$

and this provides a method of calculating entropies by means of specific heat determinations. Figure 2.7 shows a typical graph of C_p plotted against $\ln T$ for a substance over a range of temperatures including a phase transformation. The jump in the curve at the transformation temperature should be noted,

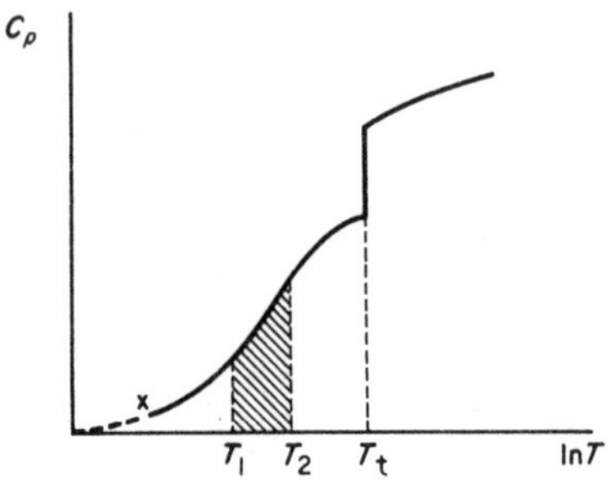

FIG. 2.7. Method of entropy determination by graphical integration of heat capacity data.

and represents the entropy of transformation L_t/T_t which must be included in ΔS. ΔS is the area beneath the curve between temperatures T_1 and T_2. C_p determinations are carried out down to the point X using techniques such as those discussed in Section 1.10, with special precautions to be observed at very low temperatures. Measurements cannot be made down to 0°K and, below the point X, values of C_p are estimated using an equation due to Debye (1912)

$$C_v = 1944 \left(\frac{T}{\theta}\right)^3 \text{ J/°K mol.} \tag{2.14}$$

θ is a constant for each substance and the relationship holds at low temperatures.[4]

Indirect methods of entropy determination depend on the relationships between entropy changes and equilibrium constants, and between entropy changes and the e.m.f.'s of certain suitable cells. (These relationships will become apparent later in this chapter and in Chapter 5.)

Deviations from the Third Law of Thermodynamics, due to incomplete ordering of a system at 0°K, have to be taken into account because it is assumed in (2.8) that $S_0 = 0$ when standard entropies of substances are being calculated.

2.6. The Driving Force behind a Chemical Reaction

A man falls to the bottom of a cliff because his potential energy is lowered in falling, and an electric current flows along a wire because it is driven by electrical energy. Why does a chemical reaction take place? The search for the driving force behind chemical reactions led immediately to the energy change which accompanies a reaction, and can be measured directly—the enthalpy change (ΔH) at constant pressure, or change in intrinsic energy (ΔU) at constant volume—and in the middle of the nineteenth century, workers such as Berthelot carried out careful determinations of enthalpy changes, assuming that these would be a measure of the driving force behind reactions. The reasoning behind this should be apparent—if a system loses energy as a result of a chemical reaction, that reaction will take place spontaneously—and the greater the quantity of heat energy lost, the greater the driving force behind the reaction.

When carbon burns in air, the reaction

$$C_S + O_{2G} = CO_{2G}$$

takes place spontaneously, with an evolution of heat energy. ΔH at 25°C is $-393{,}500$ J/mol, which is a driving force of satisfactory proportions according to this concept.

$$2Fe_L + O_{2G} = 2FeO_L \quad \text{and} \quad 3FeO_L + 2Al_L = 3Fe_L + Al_2O_{3_S}$$

both have negative heats of reaction at 1600°C, and both take place spontaneously, so that the idea still works. On the other hand, if we consider the reaction discussed in Section 1.9,

$$ZnO_S + C_S = Zn_G + CO_G.$$

ΔH at 1100°C was calculated, and found to be +348,500 J/mol, compared with +237,700 J/mol at 25°C. The reaction does not take place at 25°C, but if the system is heated to 1100°C, carbon will reduce zinc oxide to produce zinc metal. This time, a reaction is made to proceed apparently by *lowering* its driving force, which should already be working in the wrong direction in any case, according to this line of reasoning.

Thus, because *endothermic reactions* (reactions accompanied by an increase in heat energy) do take place spontaneously in addition to exothermic reactions, we cannot use heats of reaction as a criterion of their tendency to take place. We must therefore search for a more consistent measure of the driving force of a reaction.

Consider the reaction

$$ZnO_S + C_S = Zn_G + CO_G.$$

again. At 1100°C, we are above the boiling point of zinc (907°C), so that the reaction takes place between the phases indicated in the equation. At 25°C, zinc is a *solid* (m.p. 420°C), so that the reaction is

$$ZnO_S + C_S = Zn_S + CO_G.$$

Remembering the discussion at the beginning of this chapter, the most obvious difference between these reactions, apart from their temperature, is the difference in the physical state of the product of the reaction, zinc. A difference in physical state means a difference in the *state of order* of a system— and consequently in the *entropy* of the system. ΔS at 25°C should be much less than at 1100°C, because at 1100°C two molecules of gas are being produced from two solid molecules, whereas at 25°C two solid molecules only produce one gaseous molecule. A solid to gas transformation means a comparatively large entropy increase (see Section 2.2), and the following figures for the reduction of zinc oxide by carbon bear this out:

$$\Delta S_{298} = 193 \text{ J/°K mol.}$$

$$\Delta S_{1373} = 285 \text{ J/°K mol.}$$

Why not use the *entropy change* of a reaction as a measure of the driving force behind the reaction? The idea is immediately contradicted if we consider the formation of the oxide of any metal, e.g.

$$Fe_S + \tfrac{1}{2}O_{2G} = FeO_S \quad \text{at} \quad 25°C,$$

$$\Delta S_{298} = -71 \text{ J/°K mol.}$$

We know that this reaction is spontaneous—an oxide film forms readily on iron at room temperature—so that a positive entropy change in the system is not the criterion of a chemical reaction. The Second Law of Thermodynamics states that a spontaneous process is always accompanied by an increase in entropy of the system *and its surroundings* (Section 2.2). In this case, though the system (iron, oxygen and iron oxide) undergoes a loss in entropy, the surroundings (some form of container) will be heated up by the heat evolved by the reaction, which is exothermic, and will therefore undergo an increase in entropy.

The surroundings receive a quantity of heat $-\Delta H$, at constant temperature and pressure, so that the entropy increase of the surroundings is $-\Delta H/T$ [see (2.7)]—the heat absorbed divided by the temperature, where ΔH is the enthalpy change of the system, and $-\Delta H$ the enthalpy change of the surroundings. If ΔS is the entropy change of the system, then the total entropy change of the system and its surroundings is $(\Delta S - \Delta H/T)$ – referring to the *system* from the point of view of the sign of the energy change. For the Second Law of Thermodynamics to be obeyed, $(\Delta S - \Delta H/T)$ must be *positive.* Because T is always positive, we can say that $(\Delta H - T\Delta S)$ must always be *negative* for a reaction to proceed spontaneously, in order that the total entropy change of the system and its surroundings can be positive when the reaction proceeds.

We will examine this factor $(\Delta H - T\Delta S)$ for the three reactions in question. For the reduction of ZnO by carbon at 25°C,

$$\Delta H_{298} - T\Delta S_{298} = +237{,}700 - 298(193) = +180{,}300 \text{ J/mol}$$

and the reaction cannot proceed because this factor is positive, and the total entropy change of the system and its surroundings is negative.

At 1100°C,

$$\Delta H_{1373} - T\Delta S_{1373} = +348{,}500 - 1373(285) = -42{,}300 \text{ J/mol}$$

This is negative, so that the reaction will take place spontaneously at 1100°C. This explains why zinc oxide smelting must be carried out at temperatures of the order of 1100°C in order that reduction of the oxide by carbon can proceed.

For the oxidation of iron at 25°C,

$$\Delta H_{298} - T\Delta S_{298} = -264{,}400 - 298(-71) = -239{,}100 \text{ J/mol.}$$

Again, this is negative, and the reaction takes place spontaneously at 25°C.

We have now found a consistent rule – which is never contravened as it is yet another statement of the Second Law of Thermodynamics. The driving force of a reaction can be calculated as $(\Delta H - T\Delta S)$; the more negative this factor, the greater the driving force, and if the factor is positive, the reaction will not proceed spontaneously.

2.7. Free Energy

The factor $(\Delta H - T\Delta S)$ discussed in the last section has dimensions of energy – because ΔH is an energy term and ΔS is the heat absorbed divided by the absolute temperature. It has been called the change in the "free energy" of the system, "free energy" being a thermodynamic function of great importance.

Consider a system undergoing a thermodynamically reversible change at constant temperature and constant *volume*. From (1.2) and the First Law of Thermodynamics,

$$\Delta U = q - w.$$

q is the heat absorbed reversibly by the system at temperature T, and w is the maximum work done by the system (see Section 1.7).

From (2.7),

$$\Delta S = \frac{q}{T}.$$

Thus, $$\Delta U = T\Delta S - w,$$

and $$-w = \Delta U - T\Delta S.$$

$-w$ is the maximum work (whether mechanical or electrical, etc.) that can be obtained from the system, and a thermodynamic function F, the "work function" or **Helmholtz free energy** (after H. von Helmholtz) can be defined such that

$$-w = \Delta U - T\Delta S = \Delta F. \tag{2.15}$$

$$F = U - TS, \tag{2.16}$$

and it is a *thermodynamic variable*, depending only on the

state of the system—not on its history—because U, T and S are all thermodynamic variables (see Section 1.4).

At constant temperature and constant *pressure*, work may be done by the system as a result of a volume change. This will be $P\Delta V$ according to (1.4), and will not be "useful" work. The useful work will then be the maximum work, $-w$, less the energy lost due to the volume change $(-P\Delta V)$.

Then we can define the "useful work" as ΔG, the **Gibbs free energy** change of the system, where, using (2.15)

$$\Delta G = -w - (-P\Delta V).$$

$$= \Delta F + P\Delta V. \qquad (2.17)$$

Now, from (2.15),

$$\Delta F = \Delta U - T\Delta S.$$

Therefore $$\Delta G = \Delta U - T\Delta S + P\Delta V.$$

But $$\Delta U + P\Delta V = \Delta H \quad \text{[from (1.9)]}$$

and $$\Delta G = \Delta H - T\Delta S. \qquad (2.18)$$

G, the Gibbs free energy of the system, is called after J. Willard Gibbs, and is a thermodynamic variable, being dependent only on the thermodynamic variables, H, T and S.

$$G = H - TS. \qquad (2.19)$$

It is the maximum work available from a system at constant pressure, other than that due to a volume change. Most metallurgical processes work at constant pressure rather than constant volume, so that we will be more concerned with G than with F.

We have already seen the fundamental importance of the factor $(\Delta H - T\Delta S)$ in Section 2.6, so that ΔG is a measure of the "driving force" behind a chemical reaction. For a spontaneous change in the system, ΔG must be *negative*—the more negative, the greater the driving force behind the change.

Note: American textbooks use the symbols A for the Helmholtz free energy, and F for the Gibbs free energy. See ref. 4, p. 185.

2.8. Some Thermodynamic Relationships Involving Gibbs Free Energy

By definition, $G = H - TS$
$= U + PV - TS.$

Differentiating,

$$dG = dU + PdV + VdP - TdS - SdT.$$

But if we assume that a reversible process is taking place involving work due only to expansion, at constant pressure,

$$dU = \delta q - PdV \quad \text{[from (1.6)]}$$

and $\delta q = TdS,$ [see (2.6)]

so that $dU = TdS - PdV,$

and
$$dG = (TdS - PdV) + PdV + VdP - TdS - SdT$$
$$= VdP - SdT.$$

But, at constant pressure, $dP = 0$, so that

$$\left(\frac{\partial G}{\partial T}\right)_P = -S. \qquad (2.20)$$

Now, at constant temperature, $dT = 0$, and

$$dG = VdP.$$

$PV = \boldsymbol{R}T$, for an ideal gas, so that

$$dG = \boldsymbol{R}T\frac{dP}{P}.$$

Integrating between limits P_A and P_B, at constant temperature,

$$\Delta G = G_B - G_A = \boldsymbol{R}T \int_{P_A}^{P_B} \frac{dP}{P}$$

$$= \boldsymbol{R}T \ln \frac{P_B}{P_A}. \qquad (2.21)$$

This corresponds to (1.11), where the maximum work done by a gas system at constant temperature when the pressure alters from P_A to P_B is

$$w = \boldsymbol{R}T \ln \frac{P_A}{P_B}.$$

Here, $\Delta G = -w$, so that, again,

$$\Delta G = -\boldsymbol{R}T \ln \frac{P_A}{P_B} = \boldsymbol{R}T \ln \frac{P_B}{P_A}.$$

Using (2.20), if G_A is the free energy of the system in its initial state and G_B the free energy in its final state when the system undergoes a change at constant pressure,

$$dG_A = -S_A dT,$$

$$dG_B = -S_B dT,$$

and $$d(G_B - G_A) = -(S_B - S_A)dT.$$

But $\Delta G = G_B - G_A$, and $\Delta S = S_B - S_A$, so that

$$d(\Delta G) = -\Delta S \,.\, dT$$

and $$\left[\frac{\partial(\Delta G)}{\partial T}\right]_P = -\Delta S. \qquad (2.22)$$

Substituting (2.22) in (2.18)

$$\begin{aligned} \Delta G &= \Delta H - T\Delta S \\ &= \Delta H + T\left[\frac{\partial(\Delta G)}{\partial T}\right]_P \end{aligned} \tag{2.23}$$

which is known as the *Gibbs–Helmholtz equation.*

Free energy changes are changes in thermodynamic variables so that they can be added and subtracted in the same way as enthalpy changes (Section 1.9). Values of ΔH and ΔS are usually tabulated for 298°K, so that ΔG_{298} can be calculated from (2.18). ΔH and ΔS vary with temperature, and this variation can be calculated from the equations

$$\Delta H_2 - \Delta H_1 = \int_{T_1}^{T_2} \Delta C_p \, . \, \mathrm{d}T. \tag{1.18}$$

and

$$\Delta S_2 - \Delta S_1 = \int_{T_1}^{T_2} \frac{\Delta C_p}{T} \, . \, \mathrm{d}T,$$

which is a direct result of (2.8). Thus,

$$\Delta G_T = \Delta H_{298} + \int_{298}^{T} \Delta C_p \, . \, \mathrm{d}T - T \, . \, \Delta S_{298} - T \int_{298}^{T} \frac{\Delta C_p}{T} \, . \, \mathrm{d}T, \tag{2.24}$$

neglecting any heats of transformation, which can be taken into account by using relationships (1.19) and (2.13). This leads to a generalized formula of the type

$$\Delta G_T = \mathrm{a} + \mathrm{b}T \log T + \mathrm{c}T^2 + \mathrm{e}T^{-1} + \mathrm{f}T \tag{2.25}$$

mentioned on page 27 of ref. 3. However, experimental errors involved in determination of the data do not often justify such complex formulae, and normally two- or three-term formulae suffice, these being of the type

$$\Delta G_{\mathrm{T}} = \mathrm{A} + \mathrm{B}T, \tag{2.26}$$

$$\Delta G_{T} = \mathrm{A} + \mathrm{B}T \log T + \mathrm{C}T, \tag{2.27}$$

used in the tables of Kubaschewski and Evans.[3]

Methods of experimental determination of free energy changes will be discussed in Section 2.14 at the end of the chapter.

2.9. Chemical Equilibrium: The Equilibrium Constant

In any chemical reaction

$$\mathrm{A} + \mathrm{B} = \mathrm{C} + \mathrm{D},$$

the reaction will proceed in the "forward" direction (left to right), and also in the "reverse" direction (right to left). If the substances A, B, C and D were placed together in a box, isolated from the chemical action of their surroundings, the forward and reverse reactions would proceed.

The **Law of Mass Action** (C. M. Guldberg and P. Waage, 1867) states that the rate of chemical reaction is proportional to the "active masses" of the reacting substances. It is now understood that, by "active masses", we can mean the *concentrations* of the reacting substances, or, in the case of gaseous reactants, their *partial pressures*, where

$$p_{\mathrm{A}} = P \,.\, x_{\mathrm{A}}, \tag{2.28}$$

p_{A} being partial pressure of A, x_{A} its mole fraction, and P the total pressure of the gaseous system. The mole fraction of a substance in a phase is the number of molecules of the substance present, N_{A}, divided by the total number of molecules present in the phase, N.

$$x_A = \frac{N_A}{N}. \tag{2.29}$$

If we have pure solids or liquids involved in a reaction, their "active masses" can be considered to be unity for the purposes of this approach.

This leads to the relationship involving generalized measurements of concentration c_A, c_B, etc.,

$$\text{rate of forward reaction} = k_1 \,.\, c_A \,.\, c_B, \tag{2.30}$$

$$\text{rate of reverse reaction} = k_2 \,.\, c_C \,.\, c_D. \tag{2.31}$$

(The significance of k_1 and k_2, the "velocity constants" of reactions, will be discussed further in Chapter 4.)

The two reactions will proceed until eventually the rate of the forward and reverse reactions are the same, provided the temperature and pressure remain unaltered. No further change in the overall composition of the system occurs, and the system is said to be in a state of *chemical equilibrium*. The two reactions continue, but at equal rates, and we have a state of dynamic equilibrium.

All other factors being equal (e.g. temperature and pressure) this state of chemical equilibrium is consistent for a given reaction, and is the same whether equilibrium is approached from the forward or the reverse side of the position of equilibrium. Left to themselves, all chemical reactions tend to *approach equilibrium*, never depart from it. This is a consequence of the Second Law of Thermodynamics, as will be seen in Section 2.12.

Now, at equilibrium, forward rate = reverse rate, so that, from (2.30), and (2.31),

$$k_1 c'_A \,.\, c'_B = k_2 c'_C \,.\, c'_D,$$

where c'_A, c'_B, etc. are the concentrations of the reactants *at equilibrium*. Thus

$$\frac{c'_C \cdot c'_D}{c'_A \cdot c'_B} = \frac{k_1}{k_2} = K_c, \tag{2.32}$$

where K_c is the *equilibrium constant* of the reaction—a constant for a given reaction at constant temperature—in terms of the *concentrations* of the reactants.

If the reactants are gaseous, then partial pressures are used instead of concentrations, and

$$K_p = \frac{p'_C \cdot p'_D}{p'_A \cdot p'_B}. \tag{2.33}$$

If two or more molecules of a substance are involved in a reaction, then their concentrations (or partial pressures in the case of gases) are raised to a power equal to the number of molecules involved in the reaction.

For example, in the reaction

$$3FeO + 2Al = 3Fe + Al_2O_3,$$

$$K_c = \frac{(c'_{Fe})^3 \cdot (c'_{Al2O3})}{(c'_{FeO})^3 \cdot (c'_{Al})^2}. \tag{2.34}$$

If c_A is the molar concentration of A—that is the number of moles per unit volume—we can calculate the relationship between K_c and K_p for a gaseous reaction

$$aA + bB = cC + dD,$$

where a, b, c and d are the number of moles of A, B, C and D respectively involved in the reaction. For 1 mole of each gas

$$c_A = \frac{1}{V} \quad \text{and} \quad p_A V = \boldsymbol{R}T.$$

Therefore $p_A = c_A \boldsymbol{R}T$.

Now, from (2.33),

$$K_p = \frac{(p'_C)^c \,.\, (p'_D)^d}{(p'_A)^a \,.\, (p'_B)^b} = \frac{(c'_C \,.\, \boldsymbol{RT})^c \,.\, (c'_D \,.\, \boldsymbol{RT})^d}{(c'_A \,.\, \boldsymbol{RT})^a \,.\, (c'_B \,.\, \boldsymbol{RT})^b}$$

$$= \frac{(c'_C)^c \cdot (c'_D)^d}{(c'_A)^a \cdot (c'_B)^b} (\boldsymbol{RT})^{\Delta n},$$

where

$$\Delta n = (\mathrm{c} + \mathrm{d}) - (\mathrm{a} + \mathrm{b}),$$

which is the difference in the number of moles of the gaseous reactants between the right- and left-hand side of the chemical equation as written.

Thus

$$K_p = K_c \,.\, (\boldsymbol{RT})^{\Delta n}. \tag{2.35}$$

Where $\Delta n = 0$, and there is no change in the number of gaseous moles as a result of the reaction, $K_p = K_c$.

Often, "mixed" equilibrium constants are used, involving concentrations and partial pressures where two phases are involved, e.g.

$$Fe_L + \tfrac{1}{2}O_{2G} = FeO_L$$

$$K = \frac{c'_{FeO}}{c'_{Fe} \cdot (p'_{O_2})^{1/2}}.$$

If one of the products is continuously removed, the reaction never reaches equilibrium – and will in fact proceed to completion. For example,

$$2MgO + Si = 2Mg + SiO_2.$$

If this reaction is carried out above the boiling point of magnesium (1105°C), the magnesium vapour can be pumped off until virtually all the magnesium oxide has been reduced.

2.10. Controlled Atmospheres

The importance of the equilibrium constant can be illustrated by considering a gaseous atmosphere of carbon monoxide and carbon dioxide in contact with iron—as might be used in the atmosphere of a steel annealing furnace.[5] The object of this atmosphere is to prevent oxidation of the steel during the heat treatment process.

The reaction to be considered is

$$CO_{2_G} + Fe_S = FeO_S + CO_G$$

$$K = \frac{c'_{FeO} \cdot p'_{CO}}{p'_{CO_2} \cdot c'_{Fe}}.$$

If we consider that Fe and FeO are pure solids, their concentrations will remain unaltered, and the Law of Mass Action allows us to put c_{FeO} and c_{Fe} equal to unity.

Then,

$$K = \frac{p'_{CO}}{p'_{CO_2}}.$$

This can be measured at various temperatures, and these are typical results:

Temperature	500°C	700°C	1000°C
K	0·83	1·43	2·50.

Now, consider a gas mixture of $p_{CO}/p_{CO_2} = 1{\cdot}5$. If we used this mixture as an atmosphere at 700°C, we would have excess carbon monoxide, because

$$\frac{p_{CO}}{p_{CO_2}} > \frac{p'_{CO}}{p'_{CO_2}}.$$

If we put a steel ingot into the furnace, the system would *approach equilibrium* by reducing any iron oxide present to iron. If we increased the temperature of the furnace to 1000°C, then

$$\frac{p_{CO}}{p_{CO_2}} < \frac{p'_{CO}}{p'_{CO_2}}$$

and in order to reach equilibrium the iron would be oxidized. For "bright annealing" therefore, it would be necessary to work with a p_{CO}/p_{CO_2} ratio greater than the equilibrium ratio at the temperature of working.

If, instead of a controlled atmosphere, we start with an atmosphere of carbon dioxide and iron it will be seen that, the higher the value of the equilibrium constant, the greater the proportion of carbon monoxide at equilibrium, and if a reaction is wanted to proceed as far in the forward direction as possible, this will be favoured by a high equilibrium constant. If the reverse reaction is required, then this will be favoured by a low equilibrium constant calculated on the basis described above.

Considerations of this type are particularly important in the study of gaseous reduction of oxides in the blast furnace extraction of metals.[6] The equilibrium ratio of p'_{CO}/p'_{CO_2} can be determined for reactions such as

$$2Fe_3O_{4_S} + CO_{2_G} = 3\,Fe_2O_{3_S} + CO_G$$
$$3FeO_S + CO_{2_G} = Fe_3O_{4_S} + CO_G$$
$$Fe_S + CO_{2_G} = FeO_S + CO_G,$$

and hence the gas composition required in the iron blast furnace stack to allow reduction of the oxides present. The equilibrium constant is temperature-dependent, so that knowledge of the temperature in the stack is important for accurate application of these data.

2.11. The Equilibrium Constant and the Stability of Compounds

Consider the values of a few equilibrium constants for the reactions in which compounds are formed from their elements at 1600°C – a typical steelmaking temperature.

1. $4Cu + O_{2_G} = 2Cu_2O \quad K_1 = \frac{(c'_{Cu_2O})^2}{(c'_{Cu})^4 \cdot p'_{O_2}} = 10^3.$

2. $2Fe + O_{2_G} = 2FeO \quad K_2 = \frac{(c'_{FeO})^2}{(c'_{Fe})^2 \cdot p'_{O_2}} = 10^8.$

3. $\frac{4}{3}Al + O_{2_G} = \frac{2}{3}Al_2O_3 \quad K_3 = \frac{(c'_{Al_2O_3})^{2/3}}{(c'_{Al})^{4/3} \cdot p'_{O_2}} = 10^{20}.$

4. $8Fe + N_{2_G} = 2Fe_4N \quad K_4 = \frac{(c'_{Fe_4N})^2}{(c'_{Fe})^8 \cdot p'_{N_2}} = 10^{-5}.$

5. $2Al + N_{2_G} = 2AlN \quad K_5 = \frac{(c'_{AlN})^2}{(c'_{Al})^2 \cdot p'_{N_2}} = 10^2.$

All these equilibrium constants are calculated for the partial pressure of the gaseous reactant in atmospheres. The values are only approximate but are sufficiently accurate for the present discussion.

Assuming that we have pure metals and pure oxides (or nitrides), we can then write down their "active masses" as unity—according to the Law of Mass Action, and can calculate the equilibrium partial pressures of oxygen and nitrogen, p'_{O_2} and p'_{N_2}, for these reactions.

Reaction	Compound formed	Equilibrium pressure at 1600°C (atm)
1.	Cu_2O	$p'_{O_2} = 10^{-3}$
2.	FeO	$p'_{O_2} = 10^{-8}$
3.	Al_2O_3	$p'_{O_2} = 10^{-20}$
4.	Fe_4N	$p'_{N_2} = 10^{5}$
5.	AlN	$p'_{N_2} = 10^{-2}$

These values of equilibrium pressures are the pressures *below* which the compounds will *dissociate*, above which they will not dissociate. p'_{O_2} and p'_{N_2} are then called the *dissociation pressures* of the relevant oxides and nitrides. We have now found a basis for causing dissociation of such compounds—

by lowering the partial pressure of the gaseous component of the compound in the atmosphere in contact with the compound. This can be achieved by placing the compound in a chamber equipped with a vacuum pump, or by passing an inert gas, such as argon, containing very small amounts of the relevant gas, over the compound. We see that, under atmospheric pressure, Fe_4N will decompose, whereas to decompose Al_2O_3, we would have to provide an atmosphere containing oxygen at less than 10^{-20} atm partial pressure—impracticable with vacuum equipment or with the purest argon available. The implications of these figures are important to the vacuum metallurgist who may be trying to remove nitrogen and oxygen from molten steel by applying a vacuum—most metal nitrides are sufficiently unstable to allow such a possibility, but most metal oxides are too stable to decompose under the pressures available in commercial plant used for steel degassing.[7]

If we consider the reaction at 1600°C between iron oxide, shown by round brackets as a slag component, and carbon dissolved in liquid iron (square brackets),

$$(FeO)_L + [C]_L = [Fe]_L + CO_G,$$

$$K = \frac{p'_{CO}}{c'_{FeO} \cdot c'_C} = 10^6 \text{ approximately.}$$

Even if the concentration of carbon dissolved in liquid was very small, p'_{CO} would be very high—and we have a very efficient method of removing oxygen from molten iron by having carbon dissolved in it. This is the basis of all steelmaking techniques at atmospheric pressure and of deoxidation in vacuum degassing of steel.

We see now that the values of the dissociation pressures—based on the equilibrium constants of the dissociation reactions—are a measure of the relative *stability* of compounds. A compound whose equilibrium constant is high for the reaction by which it is formed is what we call a "stable" compound

—one whose equilibrium constant for its reaction of formation is relatively low is a relatively "unstable" compound. It should be remembered that, in the reaction

$$FeO+C=Fe+CO$$

at 1600°C, the fact that $K=10^6$ means that, at equilibrium, not all FeO will be reduced to iron, but the concentrations of FeO and C will be very small indeed, compared with those of Fe and CO. If the equilibrium constant is very small, as in the reaction

$$8Fe+N_2=2Fe_4N,$$

where $K=10^{-5}$ at 1600°C, the concentration of the product, Fe_4N, at equilibrium will be very small, but the reaction will not be a complete dissociation. When the equilibrium constant is either very small or very large, the accuracy of experimental analysis may be such that, to the nearest approximation, the position of equilibrium will be so far in the forward or the reverse direction that the reaction can be considered "complete". Nevertheless, the fundamental idea of chemical equilibrium denies the possibility of a "complete" reaction unless one of the products is completely removed from the system—as in the case of a gaseous product pumped away from the reaction chamber. At equilibrium, there will always be a finite concentration of products and reactants, no matter how small this may be.

The measurement of stability of compounds using the equilibrium constant is somewhat cumbersome, and we will see in the next section the relationship between the equilibrium constant and the free energy change accompanying a reaction. Free energy changes are a measure of the "driving force" behind a reaction (Section 2.7), and the driving force behind the reaction by which a compound is formed should be a measure of the stability of that compound.

2.12. The Free Energy Change of a Reaction in Terms of the Concentrations of the Reactants and Products of the Reaction: The van't Hoff Isotherm

In the reaction

$$A+B=C+D,$$

the free energy change accompanying the reaction will depend on the concentration of the reactants used, and on the concentration of the products. We might have pure A, or it might be dissolved in dilute solution as the component of a slag. D might be a pure metal, or it might be a minor constituent of a molten alloy. The free energy change is not the same in each case, and the approach of J. H. van't Hoff (1886) is appropriate here.

We will consider this chemical reaction involving the ideal gases A, B, C and D, where A and B are the "reactants", C and D the "products" of the reaction, which is carried out *at constant temperature T*. Initially, we have one mole each of A and B at partial pressures p_A and p_B respectively. We now change their partial pressures to the *equilibrium* values p'_A and p'_B respectively—by an isothermal reversible change. The gases are transferred into a large "box" containing A, B, C and D at equilibrium—therefore at equilibrium partial pressures, p'_A, p'_B, p'_C and p'_D respectively. The box contains a large mass of gas, and the addition of 1 mole of A and B does not alter the partial pressures in the box.

The free energy change involved in changing the partial pressures p_A and p_B to p'_A and p'_B respectively is, using (2.21),

$$\Delta G_1 = RT \ln \frac{p'_A}{p_A} + RT \ln \frac{p'_B}{p_B}.$$

As the system is at equilibrium, there is no free energy change when the gases are introduced into the "equilibrium box".

Now we take 1 mole each of the gases C and D out of the box, and change their partial pressures from p'_C and p'_D to

p_C and p_D respectively. The free energy change involved will be

$$\Delta G_2 = RT \ln \frac{p_C}{p'_C} + RT \ln \frac{p_D}{p'_D}.$$

We have now caused a net change in which 1 mole of A and 1 mole of B at p_A and p_B respectively have reacted to form 1 mole of C and 1 mole of D at p_C and p_D respectively. The total free energy change involved in the reaction is

$$\Delta G = \Delta G_1 + \Delta G_2 = RT \ln \frac{p'_A}{p_A} + RT \ln \frac{p'_B}{p_B} + RT \ln \frac{p_C}{p'_C} + RT \ln \frac{p_D}{p'_D}.$$

Rearranging this expression, and using (2.33)

$$\Delta G = -RT \ln \frac{p'_C \cdot p'_D}{p'_A \cdot p'_B} + RT \ln \frac{p_C \cdot p_D}{p_A \cdot p_B}$$

$$= -RT \ln K_p + RT \ln \frac{p_C \cdot p_D}{p_A \cdot p_B}, \qquad (2.36)$$

which is known as the *van't Hoff isotherm.*

If the *concentrations* of the reactants are used – as in the case of solids and liquids – rather than partial pressures, using a similar argument,

$$\Delta G = -RT \ln K_c + RT \ln \frac{c_C \cdot c_D}{c_A \cdot c_B}. \qquad (2.37)$$

Now, considering (2.36) again, if p_A, p_B, p_C and p_D are equal to 1 atm pressure – that is the reactants and products are in their "standard states",

$$\ln \frac{p_C \cdot p_D}{p_A \cdot p_B} = 0,$$

and

$$\Delta G = -RT \ln k_p = \Delta G^o, \qquad (2.38)$$

where ΔG° is the "standard free energy change" of the reaction. Similarly, with reactant and product *concentrations* used (2.37),

$$\Delta G = -RT \ln K_c = \Delta G^{\circ}, \tag{2.39}$$

if c_A, c_B, c_C and c_D are at unit concentration.

The standard free energy change of a reaction is an important property of the reaction – and can be useful provided the "standard states" of the reactants and products are clearly indicated. ΔG° is usually tabulated in sources such as ref. 3 for a number of important metallurgical reactions, and expression (2.36) can be used to calculate the free energy change involved in the reaction under the conditions being considered.

For example, in the reaction

$$Al_2O_{3_S} + 3C_S = 2Al_L + 3CO_G \quad \text{at } 1500°C,$$

$$Al_2O_{3_S} = 2Al_L + \tfrac{3}{2}O_{2_G}, \quad \Delta G^{\circ}_{1773} = +1{,}017{,}000 \text{ J},$$

$$3C_S + \tfrac{3}{2}O_{2_G} = 3CO_G, \quad \Delta G^{\circ}_{1773} = -866{,}000 \text{ J}.$$

Adding these two equations,

$$Al_2O_{3_S} + 3C_S = 2Al_L + 3CO_G, \quad \Delta G^{\circ}_{1773} = +1{,}017{,}000 - 866{,}000$$

$$= +151{,}000 \text{ J}.$$

But, from (2.36),

$$\Delta G_{1773} = \Delta G^{\circ}_{1773} + RT \ln \frac{p^3_{CO} \cdot c^2_{Al}}{c_{Al_2O_3} \cdot c_C{}^3}.$$

Using reactants in their standard states to give products in their standard states, there is clearly no possibility of a spontaneous change because ΔG is positive (see Section 2.7).

However, if we reduce the partial pressure of carbon monoxide gas in the system leaving the other reactants and products in their standard states – for example, by placing the reactants in a vacuum unit – the value of ΔG will *decrease* because the right-hand term will decrease. Using

$$\Delta G_{1773} = +151{,}000 + 1773\boldsymbol{R} \ln p^3_{CO},$$

ΔG can be calculated at various pressures of carbon monoxide.

p_{CO} (atm)	ΔG_{1773} (J/mol)
1	+ 151,000
10^{-1}	+ 49,000
10^{-2}	− 52,700
10^{-3}	− 154,400
10^{-4}	− 256,100
10^{-5}	− 357,700

We see now that the reaction becomes thermodynamically possible if the value of p_{CO} is low enough. We saw the effect of lowering the pressure of carbon monoxide on the reaction

$$(FeO)_L + [C]_L = [Fe]_L + CO_G$$

in Section 2.11 – from the point of view of the equilibrium constant – and this is an alternative approach, but a more convenient one. In (2.36), if we make

$$\frac{p_C \cdot p_D}{p_A \cdot p_B} < \frac{p'_C \cdot p'_D}{p'_A \cdot p'_B},$$

the equilibrium constant, then ΔG must be negative, and the reaction will proceed from left to right. This is the same thing as saying that all systems *tend to equilibrium*, because p_C, p_D, p_A and p_B will adjust themselves until

$$\frac{p_C \cdot p_D}{p_A \cdot p_B} = \frac{p'_C \cdot p'_D}{p'_A \cdot p'_B},$$

then $\Delta G = 0$, and the system will be at equilibrium.

If $$\frac{p_C \cdot p_D}{p_A \cdot p_B} > \frac{p'_C \cdot p'_D}{p'_A \cdot p'_B},$$

then ΔG will be positive, and the reaction will proceed towards equilibrium from right to left.

Using the relationship (2.38),

$$\Delta G^\circ = -RT \ln K_p,$$

for the reactions discussed in Section 2.11 in which certain metallic oxides and nitrides were formed at 1600°C, we can use the value of K_p to determine ΔG°_{1873}, the standard free energy of formation of the compounds.

Compound formed	K_p	ΔG°_{1873} J/mol of gaseous reactant	Dissociation pressure* (atm)
Fe_4N	10^{-5}	+178,900	10^5
AlN	10^2	− 71,600	10^{-2}
Cu_2O	10^3	−107,300	10^{-3}
FeO	10^8	−286,200	10^{-8}
Al_2O_3	10^{20}	−715,500	10^{-20}

(These figures are approximate, and tables such as those in ref. 3 should be consulted for more accurate data.)

Thus we can see that ΔG°, the standard free energy of formation of the various compounds, is a measure of the *relative stability* of the compounds. Al_2O_3 is the most stable, Fe_4N the least stable, and the more negative the value of ΔG°, the more stable the compound. In Section 2.11 we saw that the stability of a compound could be measured by means of its dissociation pressure—or by the equilibrium constant of the reaction by which it is formed. The van't Hoff isotherm establishes the identity between these arguments.

2.13. The Relationship between the Equilibrium Constant and the Temperature of a Reaction: The van't Hoff Isochore and the Clausius–Clapeyron Equation

If (2.38),

$$\Delta G^{\circ} = -RT \ln K_p,$$

is differentiated with respect to temperature, at constant pressure,

$$\left[\frac{\partial(\Delta G^{\circ})}{\partial T}\right]_p = -R \ln K_p - RT\left(\frac{\partial \ln K_p}{\partial T}\right)_p.$$

Multiplying through by T,

$$T\left[\frac{\partial(\Delta G^{\circ})}{\partial T}\right]_p = -RT \ln K_p - RT^2 . \frac{\mathrm{d} \ln K_p}{\mathrm{d}T}$$

$$= \Delta G^{\circ} - RT^2 . \frac{\mathrm{d} \ln K_p}{\mathrm{d}T} \quad \text{[from (2.38)].}$$

But the Gibbs–Helmholtz equation (2.23) states that

$$\Delta G = \Delta H + T\left[\frac{\partial(\Delta G)}{\partial T}\right]_p.$$

If reactants and products are in their standard states,

$$\Delta G^{\circ} = \Delta H^{\circ} + T\left[\frac{\partial(\Delta G^{\circ})}{\partial T}\right]_p,$$

and we can substitute in the equation above [derived by differentiating (2.38)],

$$T\left[\frac{\partial(\Delta G^{\circ})}{\partial T}\right] = \Delta G^{\circ} - \Delta H^{\circ} = \Delta G^{\circ} - RT^2 . \frac{\mathrm{d} \ln K_p}{\mathrm{d}T}.$$

Therefore,

$$\Delta H^{\circ} = RT^2 \, . \, \frac{\mathrm{d} \ln K_p}{\mathrm{d}T}$$

and

$$\frac{\mathrm{d} \ln K_p}{\mathrm{d}T} = \frac{\Delta H^{\circ}}{RT^2}, \tag{2.40}$$

which is known as *van't Hoff's isochore*, and shows the effect of temperature on the equilibrium constant of a reaction. "Isochore" means equal volume, so that this is misleading— it stems from van't Hoff's original deduction which was for constant volume.

Integrating (2.40),

$$\ln K_p = -\frac{\Delta H^{\circ}}{RT} + \text{constant}, \tag{2.41}$$

so that if we know K_p at one temperature, and ΔH°, we can calculate K_p at a second temperature. We have seen (in Section 1.8) that ΔH is temperature-dependent, but if the range of temperature over which the calculation is carried out is no more than about 200°C, the error introduced into the calculated value of K_p is small. If the temperature coefficient of the equilibrium constant is known, this equation allows a calculation of ΔH°.

If we have an exothermic reaction, ΔH° will be negative, and therefore K_p will decrease as the temperature increases. For an endothermic reaction, K_p will increase as the temperature increases.

If we consider the vaporization of a liquid A,

$$A_L = A_G,$$

ΔH° for this reaction is the latent heat of evaporation of A,

L_e. The equilibrium constant for this reaction will be

$$K_p = p_A^o$$

because the liquid will be in its standard state. p_A^o is the vapour pressure of pure liquid A, and from (2.40),

$$\frac{d \ln p_A^o}{dT} = \frac{L_e}{RT^2}, \tag{2.42}$$

which is known as the *Clausius–Clapeyron equation* (B. P. E. Clapeyron, 1834, and R. Clausius, 1850). This equation can be used to calculate L_e by measuring the vapour pressure of a substance at various temperatures, and is important in vacuum metallurgy where the vapour pressure of certain elements can be high enough to cause serious evaporation losses in vacuum melting and casting.

The latent heat of evaporation of manganese at its boiling point (2095°C) is 226 kJ/mol. We can therefore calculate its vapour pressure at the temperature of molten steel (1600°C).

Using the integrated form of (2.42),

$$\ln p_{1873}^o - \ln p_{2368}^o = -\frac{L_e}{R}\left(\frac{1}{1873} - \frac{1}{2368}\right).$$

$p_{2368}^o = 1$ atm, as 2368°K is the boiling point of manganese.

$$\therefore \ln p_{1873}^o = -\frac{226{,}000}{8{\cdot}314}\left(\frac{1}{1873} - \frac{1}{2368}\right)$$

$$= -3{\cdot}04$$

$$\therefore p_{1873}^o = 0{\cdot}048 \text{ atm.}$$

It must be remembered that we have assumed that L_e is constant over the range 1873–2368°K – more accurate data can be obtained from ref. 3, which expresses vapour pressures in terms of the absolute temperature by taking this into account. In addition, p°_{Mn} is the vapour pressure of *pure* manganese, not the vapour pressure exerted by a dilute solution of manganese in iron – which is of the order of 7×10^{-5} atm (see Chapter 3 on Solutions). Nevertheless, a vapour pressure of this magnitude leads to appreciable losses of manganese when molten steel is degassed at low pressures.

The Clausius–Clapeyron equation (2.42) can be extended to cover sublimation of A,

$$A_S = A_G,$$

where L_s is the latent heat of sublimation (per mole) of A, so that, using the same argument as for (2.42),

$$\frac{d \ln p^{\circ}_S}{dT} = \frac{L_s}{RT^2}, \qquad (2.43)$$

where p°_S is the vapour pressure of the pure solid A in contrast with p°_A, the vapour pressure of pure liquid A. If the liquid is supercooled to a temperature T, at which the solid is stable, subtracting (2.42) from (2.43),

$$\frac{d \ln (p^{\circ}_S/p^{\circ}_A)}{dT} = \frac{L_s - L_e}{RT^2} = \frac{L_f}{RT^2}, \qquad (2.44)$$

where L_f is the latent heat of fusion of A.

If A is a component of a liquid solution, whose solute is *insoluble in solid* A, at the liquidus temperature of that solution (the temperature at which solid A first appears under equilibrium conditions), then the value of p°_S will be the vapour pressure of the solution of A at that temperature. Then $p^{\circ}_S/p^{\circ}_A = x_A$, the mole fraction of A in the liquid solution, and

$$\frac{d \ln x_A}{dT} = \frac{L_f}{RT^2}. \qquad (2.45)$$

When $x_A = 1$ (pure A), the liquidus temperature is the freezing point of pure A, T_f, so that we can integrate between T_f and T_L, liquidus temperature of the solution,

$$\ln x_A = -\frac{L_f}{R}\left(\frac{1}{T_L} - \frac{1}{T_f}\right). \tag{2.46}$$

This equation can be used to calculate T_L – the liquidus temperature of the solution, and hence $T_f - T_L$, the depression of the freezing point of A. It must be emphasized that this application is only valid if there is no appreciable solid solubility – such as is found in the systems Cd–Zn, Cd–Bi. The equation can also be used to calculate L_f, knowing values of T_L and T_f. A good correlation between experimental and calculated results is in general only found with reasonably dilute liquid solutions.

2.14. Experimental Determination of Free Energy Changes

The free energy change of a reaction can be calculated from expressions such as (2.18),

$$\Delta G = \Delta H - T\Delta S,$$

by determining ΔH and ΔS (see Sections 1.10 and 2.5 respectively).

$$\Delta G^{\circ} = -RT \ln K_p,$$

so that the standard free energy of reaction can be calculated if the equilibrium constant of the reaction is determined experimentally. This involves determination of the concentrations of reactants and products at equilibrium in a constant-temperature system. Knowing ΔG°, ΔG for the conditions under which a reaction is to take place can be calculated from (2.36).

It will be seen in Chapter 5 that the free energy of a reaction in a reversible cell is related to the electromotive force of that cell, and this may form a useful basis for free energy determinations if a suitable reversible cell can be constructed.

Like ΔH, and ΔS, values of ΔG can be added to and subtracted from one another, but again there is a danger of introducing excessive errors (see Section 1.9 referring to Hess's law computations). References 3, 8, 9 and 10 may be consulted for details of experimental techniques and for suitable thermodynamic data.

References

1. WILKS, J. *The Third Law of Thermodynamics*, Oxford University Press, 1961.
2. BUTLER, J. A. V. *Chemical Thermodynamics*, Macmillan, London, 1951. Chapter XXII by W. J. C. ORR, The Application of Statistical Mechanics to the Determination of Thermodynamic Quantities.
3. KUBASCHEWSKI, O. and EVANS, E. LL. *Metallurgical Thermochemistry*, Pergamon, London, 1958.
4. DARKEN, L. S. and GURRY, R. W. *Physical Chemistry of Metals*, McGraw-Hill, New York, 1953, p. 152.
5. JENKINS, I. *Controlled Atmospheres for the Heat Treatment of Metals*, Chapman & Hall, London, 1946, p. 206.
6. WARD, R. G. *An Introduction to the Physical Chemistry of Iron and Steel Making*, Edward Arnold, London, 1962, p. 197.
7. WINKLER, O. The Theory and Practice of Vacuum Melting. *Metallurgical Reviews of the Institute of Metals*, London, **5**, No. 17 (1960), 5, 6.
8. BOCKRIS, J. O., WHITE, J. L. and MACKENZIE, J. D. *Physico-chemical Measurements at High Temperatures*, Butterworths, London, 1959.
9. KELLEY, K. K. *Contributions to Data on Theoretical Metallurgy*, U.S. Bureau of Mines, U.S. Government Printing Office, Washington, Bulletin 383, 1935; Bulletin 584, 1960; Bulletin 592, 1961.
10. ELLIOTT, J. F. and GLEISER, M. *Thermochemistry for Steelmaking*, Addison-Wesley, Reading, Massachusetts, Vol. I, 1960; (and RAMAKRISHNA, V.) Vol. II, 1963.

CHAPTER 3

Solutions

3.1. Introduction

The Law of Mass Action, which was discussed in Section 2.9, stated that the rate of chemical reaction is proportional to the "active masses" of the reacting substances. The term "active mass" refers to the amount of substance which is active or available in the reaction, and can be taken as the molecular concentration of the substance present. In practice, substances are not pure, and exist in mixtures or solutions. A solution, whether it be gaseous, liquid or solid, is a homogeneous mixture of two or more substances; it is a single phase. The active masses of reactants must therefore be related to the molecular concentrations of the reactants in the solutions in which they exist. In the reaction of a corrosive medium with an alloy of copper or zinc, the active mass of the copper would be related to its atomic concentration in the alloy, and in a slag-metal reaction, the active mass of one of the slag components—silica for instance—would be related to its concentration in the slag.

The above approach is justified under ideal conditions, which assume that the components of a solution act independently of one another, but what of the situation where the components of a solution *interact* in some way with one another? If lime is added to the slag mentioned above, there will be a strong interaction between the calcium ions and oxygen ions introduced, and the silica already present. Such an interaction would reduce the availability of the silica for other reactions, and the rate of these reactions would no longer depend on the concentration of the silica in the slag, but on a certain fraction of that concentration, called the *activity* of the silica in the slag. This activity of a substance might be expected to vary with the composition of the solution and perhaps with the temperature.

When two components of a solution are mixed, there will be a free energy change and an entropy change involved, resulting in either an evolution or an absorption of heat by the system. Such changes can be measured, and can lead to a thermodynamic approach to the study of solutions such as those in alloy systems, slags and aqueous systems.

This chapter will consider the concept of activities in solution and the thermodynamics of solutions, which are of fundamental importance to the student of metallurgy—whether his interest is in metal extraction, corrosion, or physical metallurgy.

3.2. Ideal Solutions: Raoult's Law

A measure of the mutual attraction of particles in a liquid is the vapour pressure of that liquid. This is the pressure of the vapour of a substance which is in equilibrium with its liquid at constant temperature. The stronger the attraction, the lower will be the pressure which the vapour in equilibrium with the liquid exerts. Approximate values for the vapour pressures of pure metals at 1600°C are given in Table 3.1. If a solute is added to the liquid, then the vapour pressure exerted by the solvent will be lowered, but the solute itself will also exert its own vapour pressure. The vapour pressure of the solution will be the sum of the separate vapour pressures of solvent and solute (see Fig. 3.1).

One form of *Raoult's law* (F. M. Raoult, 1887) states that the relative lowering of the vapour pressure of a solvent due to the addition of a solute is equal to the molecular fraction of the solute in the solution. Imagine two substances, A and B, forming a solution. Each substance exerts its own vapour

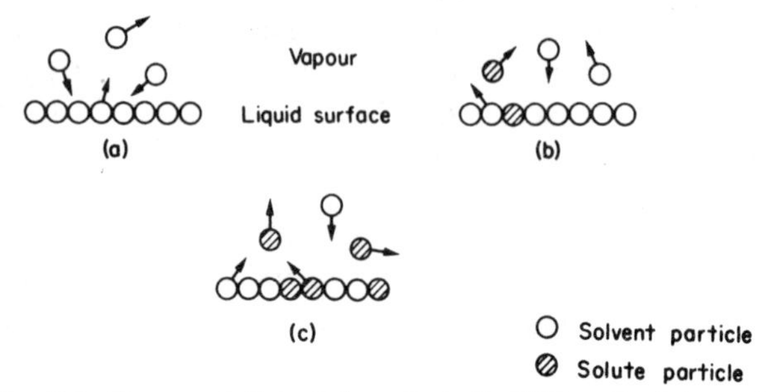

FIG. 3.1. Representation of the effect of addition of solute on the vapour pressure exerted by the solvent: (a) pure solvent in equilibrium with its vapour; (b) addition of solute lowers vapour pressure of solvent; (c) further addition of solute lowers vapour pressure of solvent, but raises vapour pressure of solute.

TABLE 3.1. APPROXIMATE VALUES FOR THE VAPOUR PRESSURES OF PURE METALS AT STEELMAKING TEMPERATURES.

Pure Metal	Approximate Vapour Pressure at 1600°C. (atmospheres)
Mg	>20
Zn	>20
Pb	0·5
Mn	0·07
Al	0·006
Cr	0·005
Cu	0·001
Fe	0·0007
Ni	0·0001

pressure, p_A and p_B respectively, at any particular composition of the solution, and the total vapour pressure (P) of the solution is equal to $p_A + p_B$. Let p_A^0 and p_B^0 be the vapour pressures exerted by the pure substances, A and B respectively; then, according to Raoult's law, if x_A and x_B are the mole fractions of A and B respectively,

$$\frac{p_A^0 - p_A}{p_A^0} = x_B \tag{3.1a}$$

and

$$\frac{p_B^0 - p_B}{p_B} = x_A. \quad (3.1b)$$

Considering (3.1a) only,

$$1 - \frac{p_A}{p_A^0} = x_B.$$

Subtracting 1 from each side, and changing the sign on each side,

$$\frac{p_A}{p_A^0} = 1 - x_B.$$

But $x_A + x_B = 1$, and therefore

$$\frac{p_A}{p_A^0} = x_A \quad (3.2a)$$

and similarly

$$\frac{p_B}{p_B^0} = x_B. \quad (3.2b)$$

This means that if a solution obeys Raoult's law the vapour pressure of one of the components of that solution is directly proportional to the mole fraction of that component in the solution. The constant of proportionality is the vapour pressure of the component in its pure state (see Fig. 3.2).

A solution which obeys Raoult's law is called an "ideal" solution. The molecules of A and B must be of similar size and must attract one another with the same force as the molecules of A attract other molecules of A, or molecules of B attract other molecules of B, and also the vapour should behave as an ideal gas for the solution to obey Raoult's law.

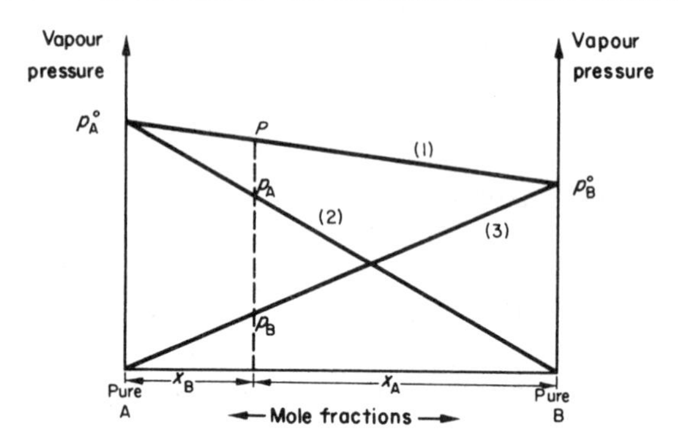

FIG. 3.2. Representation of Raoult's law for a binary solution of A and B. The straight line (1) shows the variation of the vapour pressure of the solution with variation in the composition of the solution. The effect of change in composition of the solution on the pressures exerted by A and B separately is shown by the straight lines (2) and (3) respectively.

3.3. Deviations from Raoult's Law

We will continue to assume that the vapour behaves as an ideal gas, and that the molecules are of similar size, but the attraction of A for B may be greater or less than the mutual attraction between A molecules or between B molecules.

If A attracts B more strongly than the mutual attraction of A or of B molecules, then in the solution of A and B, neither A nor B will tend to leave the liquid as readily as would be expected from an ideal solution, and the vapour pressure exerted by each component of the solution will be *lower* than that expected if the solution was ideal. This leads to a *negative* deviation from Raoult's law (Fig. 3.3).

If the attraction between A and B molecules is weaker than the mutual attraction of A molecules or B molecules, then in solution the molecules will leave the surface of the liquid more easily than would be expected for an ideal solution. The vapour pressure will be *higher* and there will be a *positive* deviation from Raoult's law (Fig. 3.4).

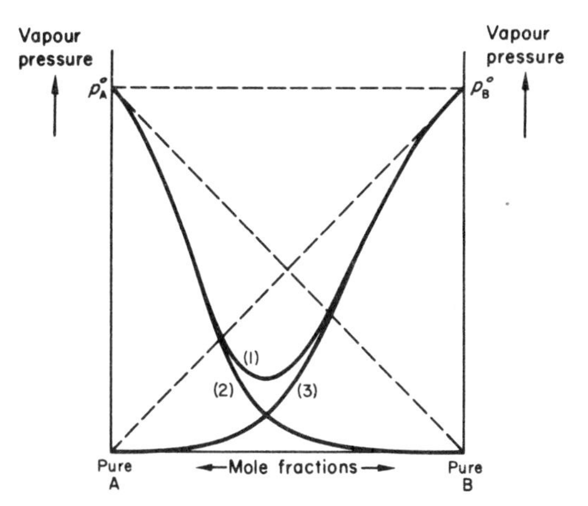

FIG. 3.3. Representation of negative deviation from Raoult's law for a binary solution of A and B. Solid curve (1) represents the vapour pressure of the solution (P), and solid curves (2) and (3) represent the vapour pressure exerted by A (p_A) and B (p_B) respectively. Dotted lines indicate the conditions for an ideal solution shown in Fig. 3.2.

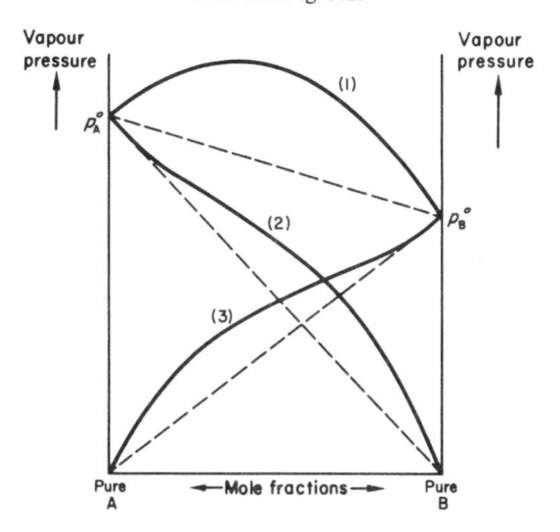

FIG. 3.4. Representation of Positive Deviation from Raoult's law, with same notation as Fig. 3.3.

An example of a solution showing a negative deviation from Raoult's law, as expressed by measurements of vapour pressures, is that between iron and nickel at 1600°C. For example, the value of p_{Ni} at 50% Ni is roughly three-quarters of its ideal value. A solution of iron and copper at 1550°C shows a positive deviation, and in a 1% copper solution in iron p_{cu} is about 10 times its ideal value. Manganese in iron is virtually an ideal solution at 1600°C.

A difference in size between the components of a solution will affect the distance between the centres of adjacent particles, and consequently the attraction between them. It would be impossible to have an ideal solution between solute and solvent particles of different sizes. To obey Raoult's law, the vapour above the solution must behave ideally, as any interaction between the particles in the vapour would affect the vapour pressure of the liquid.

If a gas does not behave ideally, we can define a quantity *fugacity* Φ so that, under all conditions,

$$\Phi V = RT.$$

In metallurgical processes, we are normally concerned with gases at low pressures or high temperatures – where effects of intermolecular attraction and the volume occupied by the molecules is insignificant – so that we can assume that the pressures exerted by gases are equal to their fugacities – that is they are ideal, and obey the Ideal Gas Equation (1.3). At high pressures or low temperatures, approaching the critical temperature of the gas, they will no longer behave ideally and it would be wrong to disregard this deviation.

3.4. Activities

It has been shown that, in actual solutions, the vapour pressure of a component is not directly proportional to the mole fraction of that component, and can either be greater or less than that expected from the solution if it obeyed Raoult's law. We can now define the "activity" a_A of the substance A in the solution so that

$$p_A = p_A^0 \, . \, a_A, \tag{3.3a}$$

and for B,

$$p_B = p_B^0 \, . \, a_B. \tag{3.3b}$$

For an *ideal* solution, it will be seen that $a_A = x_A$, the mole fraction of A in the solution, but if the solution deviates from Raoult's law,

$$a_A = \gamma_A \, . \, x_A, \tag{3.4}$$

where γ_A is a fraction (greater than unity for a positive deviation, less than unity for a negative deviation) called the *Raoultian activity coefficient.* For a pure substance, $\gamma_A = 1$ and $x_A = 1$, so that we have unit activity of the substance A, which is said to be in its "standard state".

If we return to our original argument (Sections 3.1 and 3.2) that the vapour pressure of a substance is a measure of its attraction to the solution in which it exists—and hence a measure of its availability for reaction, perhaps with another phase—we can state that this fundamental definition of activity, as that fraction of the molar concentration "available" for reaction, is universally applicable.

By definition, where the activity of a substance is changed from a_1 to a_2, the free energy change involved is

$$\Delta G = RT \ln \frac{a_2}{a_1}. \tag{3.5}$$

It should be noted that the activity coefficient of a substance in solution can vary with the concentration of that substance in the solution, and the graph showing the relationship between activity and concentration is not usually a straight line (Fig. 3.5).

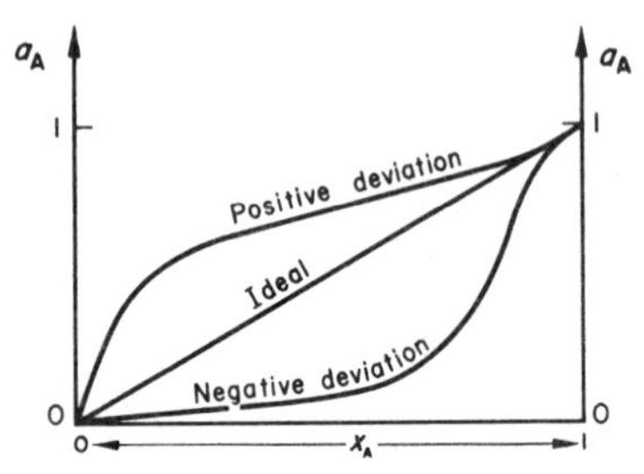

FIG. 3.5. Graph showing relationship between activity and mole fraction of substance A in three solutions—one ideal, the second showing a positive deviation from ideal behaviour, the third a negative deviation.

If the phase diagram of an alloy system is examined, it will often indicate whether a negative or a positive deviation from Raoult's law is to be expected. If a compound is formed in the solid state, there should be a negative deviation because of the strong attraction between the two components. Liquid immiscibility indicates a positive deviation because the attraction between the two components is not as strong as the bonds between the atoms of the same component, and the system must be split up into two separate phases, each containing a large proportion of one component. The structure of the liquid state can be similar to the solid state, and indications of deviation from ideal behaviour offered by the phase diagram for the solid state should be extended, to a lesser degree perhaps, into the liquid state. The converse is true, and measurements of deviation from ideal behaviour in the liquid state and at high temperatures in the solid state have been used to predict equilibrium conditions at lower temperatures in the solid state. Kubaschewski and Chart[17] discuss the chromium–molybdenum system in which a positive deviation indicated a miscibility gap below 750°C. The rate of separation at low temperatures in the solid state is so slow that annealing times of the order of a year were necessary to demonstrate the existence of the miscibility gap.

We have shown that activities are a measure of the *bonding* between constituents of a solution—so that if one of the constituents takes part in a *chemical reaction*, for example, silicon dissolved in liquid iron reacting with oxygen to form silica (a slag component),

$$[Si]_L + O_{2_G} = (SiO_2)_L,$$

the silicon does not behave as if it were pure silicon dissolved in ideal solution in iron because the Fe—Si bond is very strong and the activity of the silicon is less than its mole fraction (a negative deviation). The rate of reaction is not proportional to $x_{Si} . p_{O_2}$, but to $a_{Si} . p_{O_2}$ (assuming oxygen to behave as an ideal gas at 1600°C). In any thermodynamic expression such as that for the equilibrium constant or the van't Hoff isotherm (2.37), activities of reactants should be used rather than concentrations unless the solution is ideal, or does not deviate significantly from an ideal solution. In the example quoted, the activity of silicon in a 1 wt.% solution in iron at 1600°C is about 1/60 of its mole fraction, so that the importance of the application of the activity concept should be clear in this case.

3.5. Henry's Law and Dilute Solutions: Changing the Standard State

Henry's law (W. Henry, 1803) states that the mass of a gas dissolved by a given volume of solvent, at constant temperature, is proportional to the pressure of the gas in equilibrium with the solution. This law is more general than Raoult's law, which states that the constant of proportionality is the vapour pressure of the pure substance.

Henry's law: $p_A \propto x_A.$

Raoult's law: $p_A = p_A^0 x_A.$ (3.2a)

It has been found that, in very dilute solutions, Henry's law is obeyed by the solute, and that the vapour pressure of the solute is proportional to its concentration (Fig. 3.6); it should

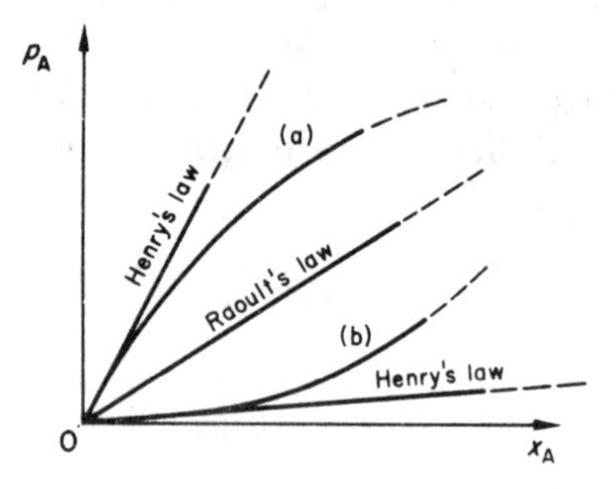

FIG. 3.6. Graph of vapour pressure of solute against mole fraction at low concentrations of solute for (a) a positive deviation from Raoult's law and (b) a negative deviation from Raoult's law. Both curves (a) and (b) approximate to a straight line at low concentrations, and are then obeying Henry's law.

be noted from this diagram that the constant of proportionality can only be p_A^0 if the solution obeys Raoult's law, as is the case with concentrated solutions, when x_A approaches unity (Fig. 3.5).

Figure 3.7 shows the relationship between the activity of the solute and the concentration when a solute obeys Henry's law.

$$a_A = \gamma_A^0 x_A \tag{3.6}$$

where γ_A^0 is the Raoultian activity coefficient of the solute A at infinite dilution—or the slope of the curve at zero concentration of A. γ_A^0 is a constant for a particular solution, and must be determined experimentally.

If we consider the solubility of the diatomic gases (O_2, N_2, H_2) in metals—we can assume that they are dissolved as single atoms and that the process of solution can be represented by

$$N_{2_G} = 2[N],$$

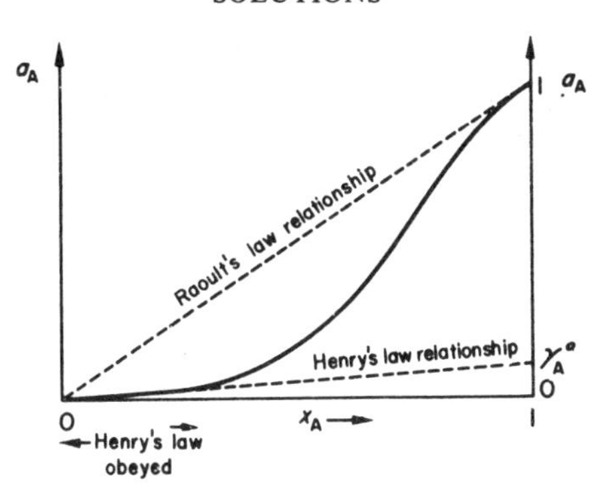

FIG. 3.7. Graph of Raoultian activity of solute A against mole fraction of A showing value of γ_A^0, the Raoultian activity coefficient of A at infinite dilution.

taking nitrogen as our example. Square brackets represent a reactant dissolved in the metal. At equilibrium, the equilibrium constant

$$K = \frac{{a'_{[N]}}^2}{p'_{N_2}}.$$

Thus $a'_{[N]} = \sqrt{(K \, . \, p'_{N_2})}$, at constant temperature, and, assuming that Henry's law is obeyed in dilute solutions, from (3.6),

$$\gamma^0_{[N]} . x'_{[N]} = \sqrt{(K \, . \, p'_{N_2})}$$

or

$$x'_{[N]} \propto \sqrt{p'_{N_2}}, \tag{3.7}$$

which is the basis of *Sieverts' law* (A. Sieverts, 1911). The solubilities of nitrogen, hydrogen and oxygen in pure liquid metals do seem to be proportional to the square roots of their partial pressures in the gas in equilibrium with the metal for dilute solutions, but the addition of other solutes can markedly effect their activities—especially in the case of oxygen—so that Sieverts' law (and therefore Henry's law) is no longer obeyed.

It is not always convenient to consider mole fractions as a measure of concentration—so that definitions of activities can be altered to accommodate this. In addition, it may be necessary to have a "standard state" other than that of the pure substance—the standard state of a substance being that in which it has *unit activity* in a particular solution. Henry's law provides the basis for an alternative definition of activities in terms of the weight per cent (wt.%) measure of concentration, where the *Henrian activity* h_A is related to the wt.% A in the solution by

$$h_A = f_A . \text{ wt.\%A}. \tag{3.8}$$

f_A is the "Henrian activity coefficient" of A, and the standard state is the 1 wt.% solution—assuming that the solution obeys Henry's law up to 1 wt.% ($f_A = 1$), then h_A is unity at this concentration. At infinite dilution, we assume that $f_A = 1$, so that as wt.%A $\rightarrow 0, f_A \rightarrow 1$ and $h_A =$ wt.%A. This system is sometimes known as the *infinite dilution* system of activities.

Deviations from Henry's law are expressed by means of Henrian activities—and the system is useful because it can often be assumed that Henry's law is obeyed in the dilute solutions encountered in steelmaking (e.g. low carbon steel in an open hearth furnace ready to tap with the composition Fe + 0·05% C, 0·05% Mn, 0·02% S, 0·01% P), and because most laboratories quote compositions in wt.% rather than mole fractions. Figure 3.8 shows the relationship between Raoultian and Henrian activities, expressed[1] by

$$f_A = \frac{h_A}{\text{wt.\% A}} = \frac{a_A}{\gamma_A^0 . x_A}. \tag{3.9}$$

It must be remembered that the fundamental thermodynamic relationships derived in Raoultian activities must be converted to the Henrian form before Henrian activities can be used in such relationships.

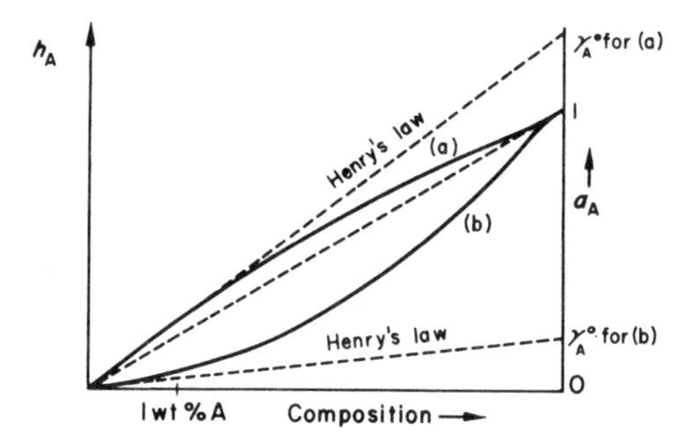

FIG. 3.8. Relationship between Raoultian activities, plotted on the right-hand vertical axis, and Henrian activities plotted on the left-hand vertical axis, for (a) a solute showing a small positive deviation from Raoult's law and (b) a solute showing a negative deviation from Raoult's law. γ_A^0 for both solutions is indicated, and the position of the 1 wt.% solution.

3.6. Experimental Determination of Activities

To illustrate this discussion on activities, and perhaps to emphasize certain points which have been made, a number of experimental methods of determination of activities in different circumstances will be outlined. (Chipman, in ref. 15, gives a detailed review of this subject for liquid metallic solutions.)

(i) *Vapour Pressure Measurements*

The measurement of the vapour pressure of one component of an alloy is a direct method of determining the activity of that component. A method involving the passing of an inert gas over the alloy and obtaining the weight of the volatil component removed in a given volume of gas, known as the "entrainment" method, has been used with alloys, and the work of Jellinek and others should be mentioned in this respect.[2] Mercury amalgams have been investigated by direct pressure measurements, for example, the study of zinc and mercury by Hildebrand.[3]

(ii) *Indirect Determination of Activity of Carbon in Liquid Steel*

The reaction $[C]+CO_{2_G} = 2CO_G$ has been used[4] as an indirect method for the determination of the activity of carbon in liquid steel. The carbon in the steel is brought to equilibrium with a carbon monoxide/carbon dioxide mixture at the temperature required and at constant pressure, and then the gas is analysed. The equilibrium constant of the reaction,

$$K = \frac{p'^{2}_{CO}}{a'_{C} \cdot p'_{CO_2}},$$

can be determined using pure carbon ($a_C = 1$) at the same temperature and pressure, and hence, knowing p'_{CO} and p'_{CO_2} in equilibrium with the steel, and by analysing the steel for its carbon content, a series of results for different carbon contents, and different temperatures, can be assembled. (*Note*: the use of partial pressures instead of activities for the gaseous reactants is justified because mixtures of gases behave as ideal solutions at the temperature of molten steel, and the fugacities of the gases are equal to their pressures.)

The results of such experiments show a negative deviation from Raoult's law, presumably because of the strong interaction between iron and carbon—although Fe_3C molecules probably do not exist in liquid steel, there is nevertheless a strong association between the iron and carbon atoms.

(iii) *Indirect Determination of Iron Oxide Activity in Molten Slags*

Chipman and others[5] determined the activity of ferrous oxide in the system FeO–CaO–SiO_2 at 1600°C by measuring the oxygen content of liquid iron in equilibrium with the slag, and comparing it with the oxygen content of liquid iron in equilibrium with pure iron oxide at the same temperature. For example, with a slag x,

$$\frac{a_{(FeO),x}}{a_{(FeO),pure}} = \frac{\text{concentration of O in metal in equilibrium with x}}{\text{concentration of O in metal in equilibrium with pure FeO}}.$$

But the activity of pure FeO slag is unity by definition, so that the activity of the FeO in the slag can be calculated.

An interesting experimental technique used was to heat the metal and slag in a crucible in a high frequency furnace, keeping the crucible spinning to prevent contact between the slag and the crucible, and consequent contamination of the slag. The metal formed a curved surface by "riding" up the sides of the crucible, and the slag (having a lower density than the metal) rested in the "dish" of molten metal.

The results were that, for a slag containing approximately two molecules of lime to one of silica, $2CaO{\cdot}SiO_2$ – the orthosilicate – the FeO activity coefficient measurements showed an increase compared with those in binary FeO–CaO melts. This is considered to be due to the strong attraction between the lime and silica components at this composition – leaving the iron oxide loosely bound, and therefore more available for reaction with the metal. For high lime content slags, a strong negative deviation was obtained, probably due to the tendency of the iron oxide to form calcium ferrite, $CaFe_2O_4$.

3.7. The Effect of Additional Solutes on the Activity of the Original Solute of a Binary Solution: Interaction Coefficients

So far, we have considered only *binary* solutions, but in metallurgical processes we are rarely dealing with binary solutions, and we must take into account the interaction between solutes in multi-component systems. If the components of binary solutions interact with one another, those of multi-component solutions will certainly interact in some way.

As an example, we will take a dilute solution of carbon in liquid iron. The Henrian activity coefficient of carbon in the binary solution f_C^C, is known[6] to be raised by the addition of small amounts of sulphur and lowered by the addition of small amounts of chromium to the solution. If we express the effect of the presence of the component i on the activity coefficient of carbon in liquid iron in the ternary Fe–C–i by f_C^i, the activity coefficient of carbon is given by

$$f_C = f_C^C \cdot f_C^i.$$

It is found that, where several components A, B, . . . , i are concerned, f_C is given approximately by the relationship

$$f_C = f_C^C \cdot f_C^A \cdot f_C^B \cdot \ldots \cdot f_C^i. \tag{3.10}$$

Thus the Henrian activity of carbon in liquid iron can be calculated by

$$h_C = f_C^C \cdot f_C^A \cdot f_C^B \cdot \ldots \cdot f_C^i \cdot \text{wt.\%C}. \tag{3.11}$$

f_C^i is the experimentally determined ratio of the activity of carbon in the Fe–C–i ternary, to the activity of carbon in the binary Fe–C,

$$f_C^i = \frac{h_C^{Fe-C-i}}{h_C^{Fe-C}}. \tag{3.12}$$

Equation (3.11) does not take into account any interactions. between the additional components (A, B, . . . , i) in the solution, so that it is only approximate where more than three components are concerned. Nevertheless, good agreement has been found between calculated and experimentally determined activities, and this approach seems reasonable.

In general, for low concentrations of B,

$$\log f_C^B = e_C^B \,.\, \text{wt.\%B}, \tag{3.13}$$

and for Raoultian activities, using the same notation,

$$\ln \gamma_C^B = \epsilon_C^B \,.\, x_B. \tag{3.14}$$

e_C^B and ϵ_C^B are called *interaction coefficients* or *interaction parameters*, and express the effect of B on C in the ternary solution in iron. Bodsworth[14] gives the relationship between these coefficients for solutions in liquid iron as

$$e_C^B = \frac{0{\cdot}2425}{M_B} \,.\, \epsilon_C^B \tag{3.15}$$

where M_B is the atomic weight of B.

Substituting (3.13) in (3.11) after taking logs (to the base 10),

$$\log h_C = \log f_C^C + e_C^A \,.\, \text{wt.\%A} + \ldots e_C^i \,.\, \text{wt.\%i} + \log \text{wt.\%C}. \tag{3.16}$$

If C interacts more strongly with B than with the iron, then e_C^B will be negative because the addition of B lowers the activity of C in the liquid iron. If, on the other hand, C interacts more strongly with the iron than with B, e_C^B will be positive, and h_C will increase as B is added.

If the coefficient e_X^Y is known, but not the coefficient e_Y^X—that is the effect of x on the activity of y in the solution being considered, then the *Wagner approximation* can be applied for very dilute solutions where the atomic weights of the solutes do not differ very much from that of the solvent; this is the approximation

$$e_Y^X = \frac{M_Y}{M_X} \,.\, e_X^Y. \tag{3.17}$$

Bodsworth[14] gives a more detailed account of this approach and tables of interaction coefficients (in an Appendix) for solutions of aluminium, carbon, chromium, cobalt, copper, manganese, nickel, nitrogen, oxygen, phosphorus, silicon, and sulphur in iron at 1600°C. Wagner[7] should be consulted for a more complete theoretical treatment.

3.8. Free Energy of Mixing

For a solution to form, the free energy change when two substances are mixed must be negative—the sign of a spontaneous process. This free energy change is known as the *free energy of mixing* ΔG^{M}. Considering the formation of a solution between two substances, A and B, where there are x_A moles of A and x_B moles of B, so that $x_A + x_B = 1$, and the activities of A and B in the solution are a_A and a_B respectively, the free energy changes *per mole of solution* when the pure substances are dissolved in the solution are

$$\Delta G_A^M = x_A \boldsymbol{R} T \ln \frac{a_A}{1}$$

and

$$\Delta G_B^M = x_B \boldsymbol{R} T \ln \frac{a_B}{1}$$

because the activities of A and B change from 1 to a_A and to a_B respectively [see (3.5)].

The total free energy change (often called the integral free energy change), the free energy of mixing, is therefore

$$\Delta G^M = \Delta G_A^M + \Delta G_B^M = x_A \boldsymbol{R} T \ln a_A + x_B \boldsymbol{R} T \ln a_B. \quad (3.18)$$

For an ideal solution, $a_A = x_A$ and $a_B = x_B$, and then the free energy of mixing is

$$\Delta G^M = x_A \boldsymbol{R} T \ln x_A + x_B \boldsymbol{R} T \ln x_B. \quad (3.19)$$

3.9. Regular Solutions

When two substances form an ideal solution, there will be no change in enthalpy. Now

$$\Delta G^{M} = \Delta H^{M} - T\Delta S^{M} \qquad \text{[see (2.18)]}$$

and, as $\Delta H^{M} = 0$,

$$\Delta G^{M} = -T\Delta S^{M} \quad \text{and} \quad \Delta S^{M} = -\frac{\Delta G^{M}}{T}$$

Using (3.19),

$$\Delta S^{M} = -x_{A}\boldsymbol{R} \ln x_{A} - x_{B}\boldsymbol{R} \ln x_{B}. \tag{3.20}$$

This is the entropy of mixing of an ideal solution—and assumes a purely random mixing of the two substances.

In the case of non-ideal solutions, it is still possible to assume random mixing in certain cases, but the enthalpy of mixing will no longer be zero because there will be heat changes due to changes in bonding energy. This assumption of random mixing can only be made where there is a small deviation from ideal behaviour, so that the enthalpy of mixing is quite small. Solutions of this type are called *regular solutions* (J. H. Hildebrand, 1927).

For regular solutions, the entropy of mixing is the same as for ideal solutions, so that, from (3.20),

$$\Delta S^{M} = -R(x_{A} \ln x_{A} + x_{B} \ln x_{B}).$$

From (3.18),

$$\Delta G^{M} = \boldsymbol{RT}(x_{A} \ln a_{A} + x_{B} \ln a_{B}),$$

and from (2.18)

$$\begin{aligned}\Delta H^{\mathrm{M}} &= \Delta G^{\mathrm{M}} + T\Delta S^{\mathrm{M}} \\ &= \boldsymbol{R}T(x_{\mathrm{A}} \ln a_{\mathrm{A}} + x_{\mathrm{B}} \ln a_{\mathrm{B}}) - \boldsymbol{R}T(x_{\mathrm{A}} \ln x_{\mathrm{A}} + x_{\mathrm{B}} \ln x_{\mathrm{B}}) \\ &= \boldsymbol{R}T\left(x_{\mathrm{A}} \ln \frac{a_{\mathrm{A}}}{x_{\mathrm{A}}} + x_{\mathrm{B}} \ln \frac{a_{\mathrm{B}}}{x_{\mathrm{B}}}\right) \\ &= \boldsymbol{R}T(x_{\mathrm{A}} \ln \gamma_{\mathrm{A}} + x_{\mathrm{B}} \ln \gamma_{\mathrm{B}}), \qquad (3.21)\end{aligned}$$

where γ_{A} and γ_{B} are the activity coefficients of A and B respectively [see (3.4)].

It is possible to use this concept of regular solutions to calculate the activities in a solution at various temperatures from measurements made at one temperature, where there is a lack of experimental data. Activity coefficients can also be estimated from known values of heats of solution. The approximate values of the thermodynamic properties of metallic systems have been obtained, but care must be taken to consider whether the solution being examined can be called a truly regular solution before the application of this technique can be fully justified. Nevertheless, thermodynamic properties calculated as for a regular solution have come close to experimentally determined data even for solutions far from regular. If there is a tendency to form compounds, or if the structure of the solution is complex (e.g. slags), this approach is unreliable.

The zinc–cadmium system has properties similar to those of regular solutions, and a detailed discussion of this system and the reason why metals do not form truly regular solutions can be found in Lumsden's *Thermodynamics of Alloys*,[8] where he uses a statistical mechanical approach not used in this book.

3.10. Partial Molar Quantities

Although the molar volume of a substance (the volume occupied by 1 mole) can be measured when the substance is in the pure state, the same amount of the substance may not take up the same volume when it is in solution. We define the *partial molar volume* of the substance, $\bar{V}$, as the volume it appears to occupy in the solution—that is, if we have a large quantity of A and B in solution, whose mole fractions are x_A and x_B respectively, and add 1 mole of A without significantly altering the composition of the solution, then the increase in volume will be $\bar{V}_A$, the partial molar volume of A. (The terms "integral" and "partial" heats of solution were introduced in Section 1.9.)

Where n_A and n_B are the number of moles present of A and B respectively,

$$x_A = \frac{n_A}{n_A + n_B}, \quad x_B = \frac{n_B}{n_A + n_B} \tag{3.22}$$

and

$$\bar{V}_A = \frac{dV}{dn_A} \tag{3.23}$$

= the rate of change of total volume with change in the number of moles of A present.

If a graph (Fig. 3.9) is drawn of the integral molar volume of the solution, V, against composition, a tangent to the curve can be drawn at any composition x_A, x_B. The partial molar volumes of A and B are then the intercepts of the tangent at $x_A = 1$ and $x_B = 1$ respectively. (V_A^o and V_B^o are the molar volumes of A and B in the pure state.)[9]

Any thermodynamic function characteristic of the state of the system which depends on the amount of substance in the system (an "extensive" property of the system) can be expressed as a partial molar quantity, and we will now consider *partial molar free energies* in the system A–B. Referring to Fig. 3.10, the integral free energy of mixing of a particular solution composition will be equal to the total change in free

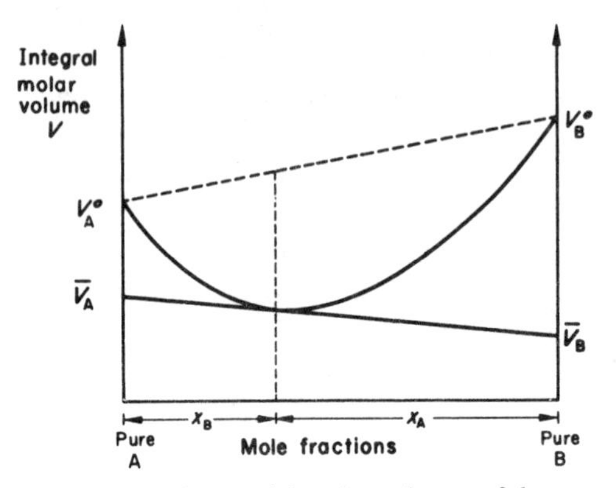

FIG. 3.9. Graph showing partial molar volumes of the components of a binary solution of A and B in relation to the integral molar volume of the solution, at constant temperature.

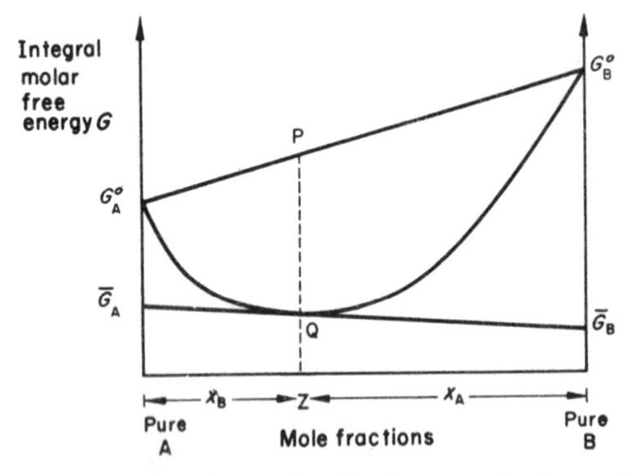

FIG. 3.10. Illustration of partial molar free energies in a solution of A and B. The free energy of mixing (ΔG^M) for the composition shown by PQZ is the distance PQ, the weighted mean of $(\bar{G}_A - G_A^\circ)$ and $(\bar{G}_B - G_B^\circ)$. The temperature is assumed constant.

energy on mixing (see Section 3.8), so that, if $(\bar{G}_A - G_A^\circ)x_A$ and $(\bar{G}_B - G_B^\circ)x_B$ are the changes in free energy of x_A moles of A and x_B moles of B respectively on entering the solution from the pure state,

$$\Delta G^{M} = (\bar{G}_{A} - G_{A}^{0})x_{A} + (\bar{G}_{B} - G_{B}^{0})x_{B}$$

$$= \Delta G_{A}^{M} + \Delta G_{B}^{M}$$

$$= \boldsymbol{R}T(x_{A} \text{ In } a_{A} + x_{B} \text{ In } a_{B}). \qquad \text{[from (3.18)].}$$

Equating coefficients,

$$\Delta G_{A}^{M} = x_{A}RT \ln a_{A}, \quad \Delta G_{B}^{M} = x_{B}RT \ln a_{B}, \qquad (3.24)$$

which corresponds with the original argument in Section 3.8. In Fig. 3.10, ΔG^{M} is the distance PQ—the weighted mean of $(\bar{G}_{A} - G_{A}^{0})$ and $(\bar{G}_{B} - G_{B}^{0})$. ($\bar{G}_{A}$, $\bar{G}_{B}$ are the partial molar free energies of A and B in solution and G_{A}^{0} and G_{B}^{0} the free energies of the pure substances A and B respectively.)

Often the partial molar free energy of a substance in solution is called the *chemical potential* μ of the substance.[10] Then $\bar{G}_{A} = \mu_{A}$, $G_{A}^{0} = \mu_{A}^{0}$, the standard chemical potential of A. Thus

$$\mu_{A} - \mu_{A}^{0} = \bar{G}_{A} - G_{A}^{0} = \boldsymbol{R}T \text{ In } a_{A}. \qquad (3.25)$$

This approach will not be used again here, as partial molar free energies seem to be more logical than chemical potentials in the context of this book.

3.11. The Gibbs–Duhem Equation

This relationship, introduced by J. Willard Gibbs (1875) and P. Duhem (1886), is used to obtain the activity coefficient of one component of a solution, knowing the activity coefficient of the other component—useful for instance in vapour pressure determinations of activities where one component is volatile, the other non-volatile (zinc–copper).

$$\Delta G^{M} = \Delta\bar{G}_{A} \,.\, x_{A} + \Delta\bar{G}_{B} \,.\, x_{B} \quad \text{(see Section 3.10)},$$

where $\Delta G_{A}^{M} = \Delta\bar{G}_{A} \,.\, x_{A}$, $\Delta G_{B}^{M} = \Delta\bar{G}_{B} \,.\, x_{B}$, $\Delta\bar{G}_{A} = \bar{G}_{A} - G_{A}^{0}$, $\Delta\bar{G}_{B} = \bar{G}_{B} - G_{B}^{0}$. Differentiating, without holding the composition constant,

$$d(\Delta G^M) = x_A \,.\, d(\Delta \bar{G}_A) + \Delta \bar{G}_A \,.\, dx_A + x_B \,.\, d(\Delta \bar{G}_B) + \Delta \bar{G}_B \,.\, dx_B.$$

But at constant composition, dx_A, dx_B and $d(\Delta G^M)$ are zero, and therefore

$$x_A \,.\, d(\Delta \bar{G}_A) + x_B \,.\, d(\Delta \bar{G}_B) = 0, \tag{3.26}$$

which is one form of the Gibbs–Duhem equation.

But $\Delta \bar{G}_A = \boldsymbol{RT} \ln a_A$, and $\Delta \bar{G}_B = \boldsymbol{RT} \ln a_B$ [from (3.24)].

Therefore $x_A \,.\, \boldsymbol{RT}\, d \ln a_A + x_B \,.\, \boldsymbol{RT}\, d \ln a_B = 0$, at constant temperature, and, dividing by $\boldsymbol{RT}$,

$$x_A \,.\, d \ln a_A + x_B \,.\, d \ln a_B = 0. \tag{3.27}$$

Thus,

$$d \ln a_B = -\frac{x_A}{x_B} d \ln a_A,$$

and, integrating between $x_B = 1$ and x_B, the value of concentration in which we are interested,

$$\ln a_B = -\int_1^{x_B} \frac{x_A}{x_B} \,.\, d \ln a_A. \tag{3.28}$$

The value of a_B can be obtained by carrying out the integration in (3.28) graphically, plotting x_A/x_B against $(-\ln a_A)$. The area under the curve between $x_B = 1$ and x_B (the value at which a_B is required) is $(-\ln a_B)$. Examination of a typical graph from such a plot (Fig. 3.11) shows that the integration is difficult because the area under the curve approaches infinity as x_B becomes small (x_A/x_B tends to infinity), and $\ln a_A$ approaches minus infinity when x_B tends to one.

However, $x_B = 1 - x_A$ for a binary mixture, so that $dx_B = -dx_A$, and

$$\frac{x_B}{x_B} \,.\, dx_B = \frac{x_A}{x_A} (-dx_a),$$

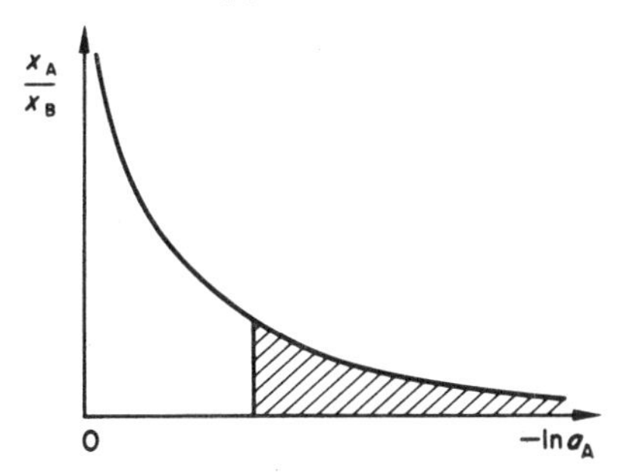

FIG. 3.11. Graphical integration of Gibbs–Duhem equation to obtain the activity of one component (a_B) knowing the activity of the other (a_A) in a binary system. The shaded area gives the value of ($-\ln a_B$) for the appropriate value of x_B.

and, as $\mathrm{d}\ln x = \dfrac{\mathrm{d}x}{x}$,

$$x_B \,.\, \mathrm{d}\ln x_B = -x_A \,.\, \mathrm{d}\ln x_A. \tag{3.29}$$

Subtracting (3.29) from (3.27),

$$x_B \,.\, \mathrm{d}\ln a_B - x_B \,.\, \mathrm{d}\ln x_B + x_A \,.\, \mathrm{d}\ln a_A - x_A \,.\, \mathrm{d}\ln x_A = 0;$$

thus

$$x_B \,.\, \mathrm{d}\ln \frac{a_B}{x_B} + x_A \,.\, \mathrm{d}\ln \frac{a_A}{x_A} = 0,$$

and from (3.4),

$$x_B \,.\, \mathrm{d}\ln \gamma_B + x_A \,.\, \mathrm{d}\ln \gamma_A = 0, \tag{3.30}$$

where γ_A and γ_B are the Raoultian activity coefficients of A and B respectively. Integrating (3.30),

$$\ln \gamma_B = -\int_1^{x_B} \frac{x_A}{x_B} \,.\, \mathrm{d}\ln \gamma_A, \tag{3.31}$$

which can be integrated graphically by plotting x_A/x_B against $\ln \gamma_A$ (Fig. 3.12). Since γ_A is always finite, there is no difficulty in integrating in the region $x_B = 1$, but there is still difficulty where x_B is small because x_A/x_B again approaches infinity.

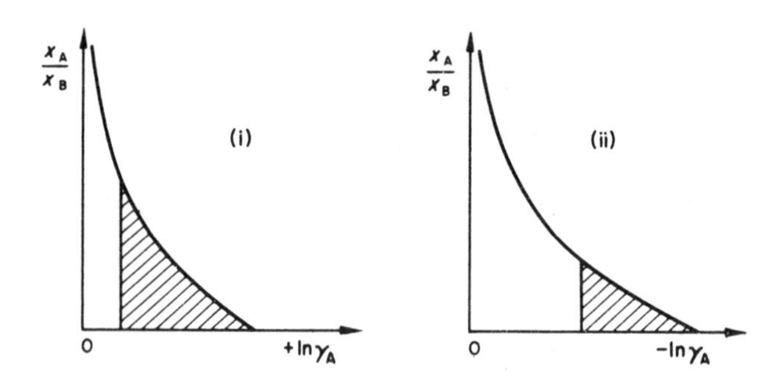

FIG. 3.12. Graphical integration of Gibbs–Duhem equation to obtain the activity coefficient of one component of a binary system (γ_B) knowing the activity coefficient of the other (γ_A). (i) shows the case of a positive deviation from ideal behaviour (γ_A and γ_B always greater than unity). The shaded area gives the value of $\ln \gamma_B$ for the appropriate value of x_B. (ii) shows the case of a negative deviation from ideal behaviour (γ_A and γ_B always less than unity). The shaded area gives the value of $(-\ln \gamma_B)$ for the appropriate value of x_B.

This problem can be overcome using special functions which cause the curve to approach the axis at a large angle and make extrapolation much more dependable.[11] Chipman, on page 38 of ref. 15, shows how (3.31) can be used for the system Fe–FeS, where γ_{Fe} is known for a range of values of x_{Fe}, and γ_{FeS} is to be calculated.

3.12. Excess Integral and Partial Molar Quantities

It is possible to divide thermodynamic properties of a solution into two parts—the part which would be exhibited by an ideal solution and the "excess" part which expresses the deviation from ideal solution.[16]

$$G = G^{I} + G^{E}, \tag{3.32}$$

where G is the integral Gibbs free energy value for the solution, G^{I} the value which G would have if the solution were ideal, and G^{E} the "excess" integral free energy – the difference between G and G^{I}. The value of the free energy of mixing, ΔG^{M}, is then given by the relation

$$\Delta G^{M} = \Delta G^{I} + G^{E}, \tag{3.33}$$

and we know from (3.19) that, for an ideal solution of A in B,

$$\Delta G^{I} = x_A \boldsymbol{R}T \ln x_A + x_B \boldsymbol{R}T \ln x_B.$$

Therefore, $$\Delta G^{M} = x_A \boldsymbol{R}T \ln x_A + x_B \boldsymbol{R}T \ln x_B + G^{E}. \tag{3.34}$$

Excess thermodynamic quantities obey all thermodynamic relationships, and we can write, from (2.19),

$$G^{E} = H^{E} - TS^{E}. \tag{3.35}$$

Using (3.20) for an ideal solution,

$$\Delta S^{I} = -x_A \boldsymbol{R} \ln x_A - x_B \boldsymbol{R} \ln x_B,$$

so that $$\Delta S^{M} = \Delta S^{I} + S^{E},$$

and $$\Delta S^{M} = -x_A \boldsymbol{R} \ln x_A - x_B \boldsymbol{R} \ln x_B + S^{E}. \tag{3.36}$$

Thus,

$$\Delta H^{M} = \Delta G^{M} + T\Delta S^{M} \quad \text{[from (2.18)]}$$

$$= x_A \boldsymbol{R}T \ln x_A + x_B \boldsymbol{R}T \ln x_B + G^{E} + T(-x_A \boldsymbol{R} \ln x_A - x_B \boldsymbol{R} \ln x_B + S^{E})$$

$$= G^{E} + TS^{E} = H^{E}. \tag{3.37}$$

For a *regular* solution, $S^{E} = 0$, and $\Delta H^{M} = G^{E}$. For an *ideal* solution, G^{E} and $S^{E} = 0$ and $\Delta H^{M} = 0$.

This alternative approach can be extended to *partial* molar quantities (Section 3.10), and we can write

$$\Delta\bar{G}_i = \Delta G_i^I + \bar{G}_i^E \tag{3.38}$$

for the component i in a solution. $\bar{G}_i^E$ is the excess partial molar free energy of i in the solution. From (3.24),

$$\Delta\bar{G}_i = \boldsymbol{R}T \ln a_i = \bar{G}_i - G_i^\circ,$$

so that

$$\Delta G_i^I = \boldsymbol{R}T \ln x_i,$$

because $\Delta\bar{G}_i = \Delta G_i^I$ and $a_i = x_i$ for an ideal solution. Therefore, using (3.38),

$$\begin{aligned}\bar{G}_i^E &= \boldsymbol{R}T \ln a_i - \boldsymbol{R}T \ln x_i \\ &= \boldsymbol{R}T \ln \frac{a_i}{x_i} \\ &= \boldsymbol{R}T \ln \gamma_i \end{aligned} \tag{3.39}$$

from the definition of activity coefficients in (3.4).

Again, using the same notation,

$$\bar{G}_i^E = \bar{H}_i^E - T\bar{S}_i^E, \tag{3.40}$$

where

$$\bar{S}_i^E = \bar{S}_i - \bar{S}_i^I. \tag{3.41}$$

It will be seen from (3.39) that the excess partial molar free energy of a component of a solution is an alternative method of expressing the activity coefficient γ_i of that component. If the solution shows a negative deviation from Raoult's law, γ_i will be less than unity and $\bar{G}_i^E$ will be negative; $\bar{G}_i^E$ will be positive for a positive deviation.

3.13. Application of Free Energy–Composition Curves to the Study of Alloy Systems

If the integral free energy is plotted for a solid solution of A in B against composition of the solution at a particular temperature, a curve such as Fig. 3.10 will be obtained. Note that, at any composition, the solid solution has lower free energy than either pure A or pure B, and will therefore be the stable state of the system at that temperature, rather than a heterogeneous mixture of A and B or of two different solutions of A and B. At the composition (x_A, x_B) shown by the vertical line PQZ, the free energy of the mixture of pure A and pure B would be shown by the point P, and that of the solid solution by the point Q – ΔG^M, the free energy of mixing, is then equal to PQ, as was mentioned in Section 3.10. This is the case for complete solubility of A in B at all compositions at the temperature shown.

If only partial solid solubility occurs at this temperature, and perhaps two solid solutions – the A-rich α, and B-rich β solid solutions – are formed, then the free energy diagram would be like Fig. 3.13. Once again PQZ represents the composition of the alloy, and P is the free energy of a mixture of pure A and pure B. R is the free energy of β solid solution of A in B, S the free energy of α solid solution of B in A. The free energy of the stable *mixture* of solid solutions α and β is shown by the point Q, where the composition line PQZ cuts the common tangent to both α and β curves, TQM. The equilibrium alloy would then consist of α of the composition shown by TE, β of composition shown by MF, present in the proportions

$$\frac{\text{number of atoms in phase } \alpha}{\text{number of atoms in phase } \beta} = \frac{\text{ZF}}{\text{EZ}} = \frac{\text{QM}}{\text{TQ}}$$

(commonly known as the "Lever rule"). It will be seen that the free energy of mixing for this alloy would be represented by the distance PQ.

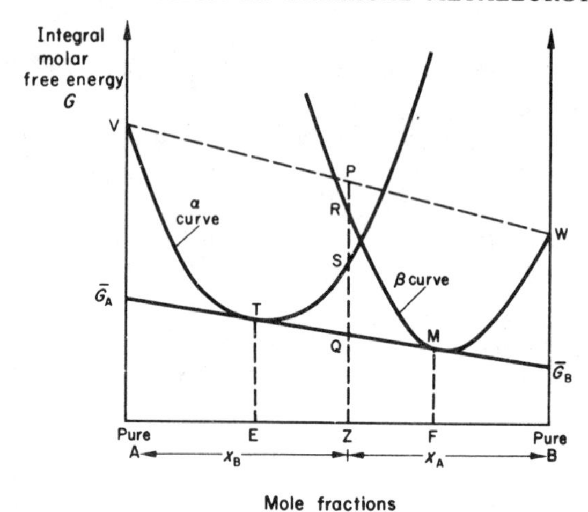

FIG. 3.13. Free energy–composition curves for the system A–B, showing partial solid solubility of A in B and B in A. The temperature is assumed constant. It should be noted that the partial molar free energy of the component A is the same in both phases present (α and β) at equilibrium, and the same applies to the component B.

At equilibrium, any composition of the alloy lying between E and F at the temperature chosen would consist of a mixture of the two solid solutions α and β, whereas any alloy of composition lying to the left of E would be of α only, and to the right of F would be of β only.

Figure 3.14 shows free energy curves for the solid and liquid phases of a eutectic type of diagram, at a number of different temperatures, and can be compared with the phase diagram shown in (vi). The free energy diagram (i) represents the state of the system at temperature T_1 in (vi), and the liquid phase is the equilibrium phase at all compositions at this temperature. As the temperature is lowered to T_2, the solid α phase becomes more stable at compositions to the left of C in (ii), and between C and D a mixture of liquid and α phases (obeying the Lever rule as in Fig. 3.13) will be the stable state. To the right of D, the alloy would still be wholly liquid at this temperature. At temperature T_3 the α and β curves both pass below the liquid

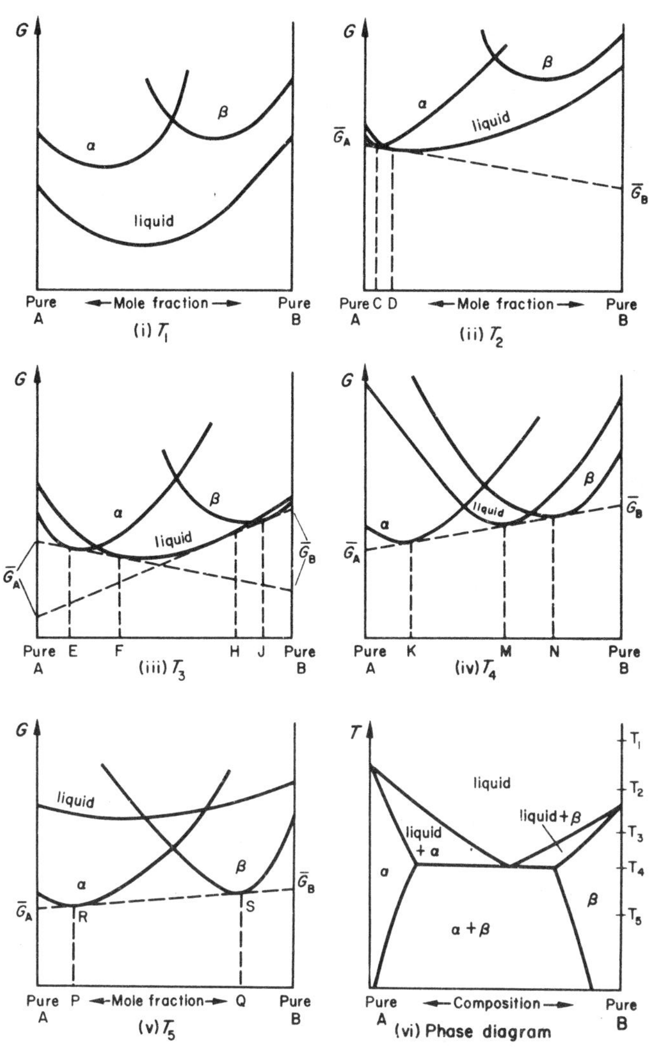

FIG. 3.14. Free energy–composition curves for the system A–B at five different temperatures, and the phase diagram of the system showing the temperatures chosen. (See text for full description.)

curve, and to the left of E and to the right of J, α and β respectively are the stable phases. Between E and F a mixture of α and liquid, between F and H liquid only, and between H and J a mixture of liquid and β will be the stable state of the alloy at this temperature. At T_4, the eutectic temperature, the common tangent to the α and β curves is also a common tangent to the liquid curve—and this is a temperature unique to the system. M represents the eutectic composition, where, at this temperature, liquid, α and β are all present in the stable state of the system. The composition of the α in the eutectic mixture is shown by K, that of the β by N. At any composition between K and M, the alloy will consist of primary α and a eutectic of α and β (the compositions of all the α and β still shown by K and N) in the equilibrium state. To the left of K and to the right of N, α and β respectively will be the stable states of the system. Diagram (v) shows the system at T_5, below the eutectic temperature—when no liquid appears in the equilibrium condition at any composition because the liquid curve no longer cuts the α curve to the left of P, nor the β curve to the right of Q, nor does it cut the common tangent to the α and β curves between R and S, the respective points of contact. Between P and Q we have mixture of α and β, and to the left of P and to the right of Q, we have only α or β respectively.

These principles can be applied to more complex systems than this, and Fig. 3.15 is shown as an example, where we have a peritectic and a eutectic reaction occurring in the alloy system A–B, separated by the intermediate phase β. Notice that if the intermediate phase is a solid solution, its free energy curve will form a broad "U", giving stability over a range of composition. If, on the other hand, it is a compound of almost stoichiometric composition, its free energy curve will be a very sharp "V", giving stability only at one composition—or rather over a very narrow range of compositions as compounds usually vary slightly from strict valency rules.

This application of the principles of the free energy of

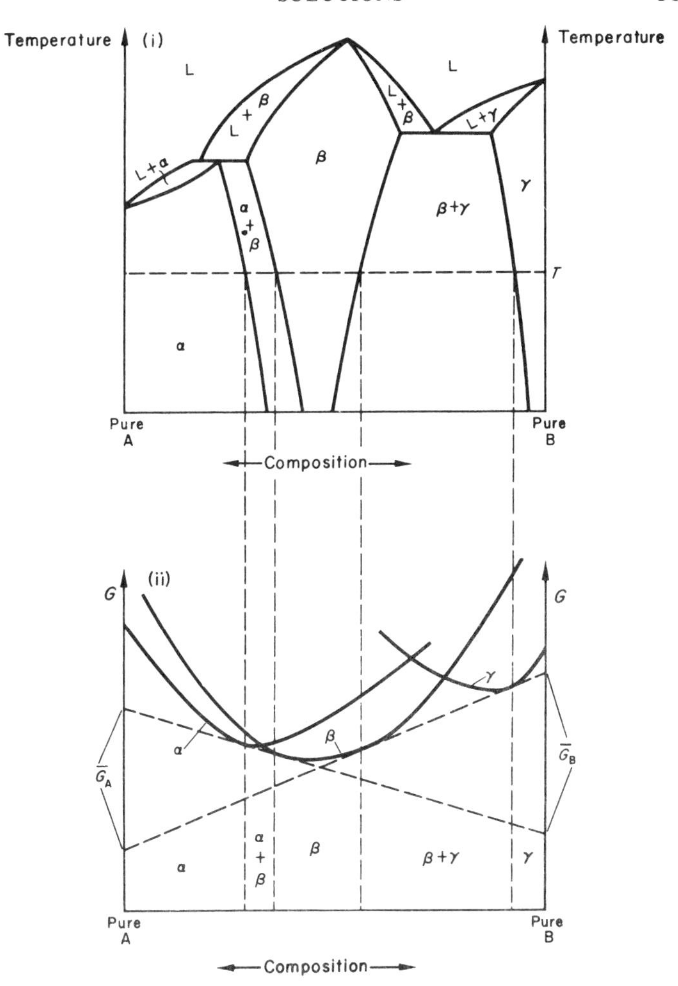

FIG. 3.15. (i) Phase diagram of an alloy system A–B, showing a peritectic and a eutectic reaction separated by an intermediate phase β. (ii) Free energy–composition diagram for the system A–B at temperature T shown in (i) above.

mixing to alloy systems is of great interest to the structural metallurgist, and the reader is recommended to consult a number of references for further discussion of the subject.[12] Kubaschewski and Chart[17] show how points on equilibrium diagrams can be calculated from thermochemical data, and how the diagrams can often be improved by combining the results of such calculations with those of the usual techniques (thermal analysis, *X*-ray diffraction studies, etc.).

References

1. WARD, R. G. *An Introduction to the Physical Chemistry of Iron and Steel Making*, Edward Arnold, London, 1962, p. 27.
2. JELLINEK, K. and ROSNER, G. A. *Z. physikal. chem.* **143** (1929), 51; **152** (1931), 67.
   JELLINEK, K. and WANNOW, H. *Z. Elektrochem.* **41** (1931), 346.
3. HILDEBRAND, J. H. *Trans. Amer. Electrochem. Soc.* **22** (1912), 319.
4. MARSHALL, S. and CHIPMAN, J. *Trans. Amer. Soc. Metals,* **30** (1942), 695: RIST, A. and CHIPMAN, J. *The Physical Chemistry of Steelmaking* (ed. J. F. ELLIOTT), Technology Press, Massachusetts Inst. Tech., and John Wiley, New York, 1958, p. 3.
5. FETTERS, K. L. and CHIPMAN, J. *Trans. A.I.M.M.E.* **145** (1941), 95: TAYLOR, C. R. and CHIPMAN, J. *Trans. A.I.M.M.E.* **154** (1943), 228.
6. FUWA, T. and CHIPMAN, J. *Trans. A.I.M.M.E.* **215** (1959), 716: PEARSON, J. and TURKDOGAN, E. T. *J. Iron and Steel Institute*, **176** (1954), 19.
7. WAGNER, C. *Thermodynamics of Alloys*, Addison Wesley, Cambridge, Mass., 1952, pp. 47–53.
8. LUMSDEN, J. *Thermodynamics of Alloys*, Institute of Metals Monograph No. 11, 1952, Chapter 18.
9. A proof of this statement, for free energy rather than volume, can be found in ref. 13, p. 241, or ref. 14, p. 39.
10. GLASSTONE, S. *Textbook of Physical Chemistry*, Macmillan, London, 1953, p. 238.
11. See ref. 13, p. 264; ref 14, p. 44; ref. 7, p. 16.
12. Ref. 13, Chapter 13; ref. 7, Chapters 4, 5, 6; and COTTRELL, A. H. *Theoretical Structural Metallurgy*, Edward Arnold, London, 2nd edn., 1955, Chapters X and XI.
13. DARKEN, L. S. and GURRY, R. W. *Physical Chemistry of Metals*, McGraw-Hill, New York, 1953.
14. BODSWORTH, C. *Physical Chemistry of Iron and Steel Manufacture*, Longmans, London, 1963, p. 73.
15. CHIPMAN, J. Activities in Liquid Metallic Solutions. Contribution to *Disc. Faraday Soc.*, No. 4 (1948). *The Physical Chemistry of Process Metallurgy*, Butterworths, London, 1961, p. 23.
16. See ref. 13, pp. 267–70.
17. KUBASCHEWSKI, O. and CHART, T. G. *J. Institute of Metals*, **93** (1964–5), 329.

CHAPTER 4

Reaction Kinetics

4.1. Introduction

It has been demonstrated that the conditions of chemical equilibrium can often be predicted by thermodynamic considerations and therefore the tendency of a reaction to proceed can be measured. The thermodynamic approach does not, however, give any indication of the rate at which a reaction will proceed, and two examples of metallurgical phenomena show the importance of the study of the mechanism and rate of a reaction. If a plain carbon steel in the austenitic state (say at 850°C) is quenched in water, a metastable martensitic structure is produced. This structure will tend to transform to a mixture of ferrite and cementite of lower free energy, but the rate of transformation at room temperature is so slow that the martensite remains apparently unchanged. "Tempering" at 650°C allows the transformation to be completed within an hour. The reaction between carbon and oxygen in liquid steel to form carbon monoxide is accompanied by a substantial negative free energy change, but the carbon and oxygen contents can be raised well above the equilibrium values without any appreciable reaction. If an inert gas such as argon is bubbled through the liquid steel, introducing suitable nuclei for reaction, a vigorous "boil" caused by carbon monoxide gas evolution can be produced. Rates vary between the extremes of the almost instantaneous neutralization of a strong acid by a strong base, and the virtually undetectable combustion of a gaseous fuel in air at room temperature, and the subject of "chemical kinetics" includes the study of the influence of variables such as temperature and concentration on reaction rates. It would be fair to say that experimental work on the kinetics of reaction – particularly in metallurgical processes – lags far behind that on thermodynamics, but it must be remembered that the experimental difficulties are much more formidable in the former field than in the latter.

Methods used to follow the rate of a reaction involve the determination of the amount of reactants remaining or products present after a given time. If possible, the method of measurement used should be applied continuously without involving a delay in the process or in the measurement, and, in this respect, physical methods (such as a continuous measurement of the conductance of an electrolyte) are to be preferred to chemical methods which involve the removal of samples for analysis. Techniques of continuous analysis of metal and slag in steelmaking processes by electrochemical methods, which are in the experimental stage, could make available a much more comprehensive volume of data on rates of reactions in these processes than has hitherto been produced. Continuous temperature measurement will also be necessary as temperature influences reaction rates considerably, and it is encouraging to hear that techniques of continuous pyrometry are being developed at the same time as continuous analysis for reactions in liquid phases at temperatures as high as 1600°C.

There are two types of reaction which will be considered: *Homogeneous* reactions take place entirely within one phase, such as reactions between gas molecules to produce gaseous products, or reactions in an aqueous solution where the reactants and products all remain dissolved in water.

Heterogeneous reactions involve more than one phase, such as the transfer of a substance from liquid slag to a liquid metal in a smelting process, or the reaction between gaseous oxygen and a solid metal to form an oxide film on the metal surface.

Homogeneous reaction mechanisms tend to be less complex than those of heterogeneous reactions, so that the preliminary theoretical treatment will be mainly that for homogeneous reactions. Metallurgical processes usually involve heterogeneous reactions, and consideration is given later in this chapter and Chapter 6 to the kinetics of these reactions.

4.2. Effect of Concentration of Reacting Substances

It was first demonstrated by L. Wilhelmy (1850) that the rate of a chemical reaction is proportional to the concentration of the reacting substances. This led to the formulation of the Law of Mass Action and the concept of the Equilibrium Constant which were discussed in Chapter 2. For the moment we will assume that the temperature, which has an important effect on reaction rates, remains constant.

We can define the *order* of the reaction as "the sum of the powers to which the concentrations of the reacting atoms or molecules must be raised to determine the rate of reaction". It must be emphasized that the order of a reaction does not necessarily bear any relation to the *molecularity* of the reaction which is "the number of atoms or molecules taking part in a reaction". For example, in the reaction

$$2HI = H_2 + I_2$$

two molecules of hydrogen iodide take part and therefore the reaction is "bimolecular". Experiments have shown that the reaction rate is proportional to the square of the hydrogen iodide concentration – indicating a "second order" reaction, whose rate depends on the concentrations of two molecules. In this case, the molecularity is the same as the order of the reaction, but where one reactant is present in excess, its concentration is not changed appreciably as the reaction proceeds, as in the reaction of hydrogen iodide with hydrogen peroxide in aqueous solution,

$$HI + H_2O_2 = HIO + H_2O,$$

where the hydrogen iodide is present in excess. The rate of this reaction depends only on the concentration of the hydrogen peroxide because the concentration of HI is not altered significantly by the reaction, and it is therefore of first order even though it is bimolecular.

In this particular case, the reaction is not completed after this first stage, and a further stage in the reaction takes place,

$$HI + HIO = H_2O + I_2,$$

giving an overall reaction

$$2HI + H_2O_2 = I_2 + 2H_2O.$$

Measurement of the rate of this overall reaction, which is often used as a convenient laboratory experiment in teaching courses, shows that it still depends only on the concentration of the hydrogen peroxide and is therefore of first order. The reason for this apparent anomaly is that the second stage occurs much more rapidly than the first stage, whose rate controls the overall rate of reaction and is known as the *rate-controlling step*. The order of the rate-controlling step determines the order of the overall reaction, but the molecularity of a reaction taking place in more than one stage has no meaning, only the molecularity of the individual stages can be determined.

Ward[(1)] studied the evaporative losses of the alloying element manganese when steel was melted under a vacuum. The process obeys first order kinetics, the rate of loss depending on the manganese content of the steel, but further analysis shows that there are several steps involved in the process. These probably include:

(a) transport of manganese atoms through the liquid iron to the metal surface;

(b) evaporation of the manganese from the liquid iron surface into the vapour phase;

(c) transport in the vapour phase away from the metal surface.

In the melting furnace used by Ward, the liquid steel was stirred well by inductive stirring so that the transport in step (a) was probably rapid except for a thin non-turbulent surface layer. Convective stirring in the vapour phase meant that the slowest part of the transport in step (c) would be in a thin stagnant layer of vapour adjacent to the metal surface. Ward concluded that, at low pressures, the rate-controlling steps were transport in the metal boundary layer and evaporation from the metal surface. At higher pressures, the rate controlling steps were transport in both metal and vapour boundary layers, the latter becoming predominant as the pressure approached atmospheric pressure. This example has been used to show the complexity of reactions in metallurgical processes, and to emphasize that an apparently simple reaction is really a combination of several steps, each of which might be rate-controlling.

It should be noted that the molecularity of a reaction must always have an integral value and will never be zero, but that the order of a reaction, as experimentally determined, can be fractional or zero. An experimental determination of the order of the reaction of ammonia on a platinum surface between 800 and 1000°C

$$2NH_3 = N_2 + 3H_2$$

shows it to be of zero order. The rate does not depend on the concentrations of the reactants in the gas phase but only on the initial concentration, because the hydrogen produced is strongly adsorbed on the surface of the platinum and retards the rate of reaction so much that changing the amount of ammonia present as the reaction proceeds does not affect the rate of reaction. In the reaction of solid copper with a gas atmosphere containing oxygen to form an oxide layer on the surface of the copper,

$$4Cu + O_2 = 2Cu_2O,$$

the concentration of the copper remains constant (pure copper) so that the reaction rate would be expected to depend on p_{O_2}, the partial pressure of oxygen in the atmosphere. This would be a first order reaction, but experimental determination of the dependence of rate on p_{O_2} shows that the rate is approximately proportional to $p_{O_2}^{1/7}$ – a fractional order of reaction. The reasons for this anomaly are to be found in the structure of the oxide film, which separates the two reactants, and the mechanism of the formation and growth of the film, which is mainly by diffusion of the cuprous ions outwards through the film rather than by oxygen diffusion inwards.

Few known reactions exceed third order and elementary reaction steps are either unimolecular or bimolecular. The chance of three molecules meeting at the same time, possessing the correct energy and orientation to react simultaneously in one step, is very unlikely. Sometimes the order of a reaction varies with time – in which case the products of the reaction are either inhibiting or enhancing the reaction.

4.3. Quantitative Relationships between Rate of Reaction and Concentration of Reactants

For the purpose of this preliminary treatment, the reactants are assumed to behave ideally in solution, and no account will be taken of the fact that in practice the activities of reactants may not be the same as their concentrations.

In a reaction of *first order*, for example

$$A = X + Y,$$

the rate of reaction is proportional to the concentration of A, and this can be expressed as

$$-\frac{dc_A}{dt} = kc_A, \tag{4.1}$$

where c_A is the concentration of A at time t and k is a constant known as the *velocity constant*, *rate constant* or *specific reaction rate* (the rate of reaction when $c_A = 1$). The reason for the negative sign is that c_A is decreasing with time as the reaction proceeds.

Equation (4.1) can be integrated after rearrangement

$$-\frac{dc_A}{c_A} = k \,.\, dt.$$

$$-\int_{c_{A,O}}^{c_A} \frac{dc_A}{c_A} = k \,.\int_0^t dt,$$ ($c_{A,O}$ = original concentration of A before reaction commenced)

and
$$\ln \frac{c_{A,O}}{c_A} = kt. \tag{4.2}$$

Alternatively, if a moles of A are present initially in unit volume and x moles have reacted in unit volume after time t, so that $c_{A,O} = a$ and $c_A = a - x$

$$\ln \frac{a}{a-x} = kt. \tag{4.3}$$

(In the example used, there will be x moles of X and x moles of Y in unit volume after time t.)

If a, x and t can be measured experimentally, a graph of $\ln(a-x)$ against t will give a straight line for a first order

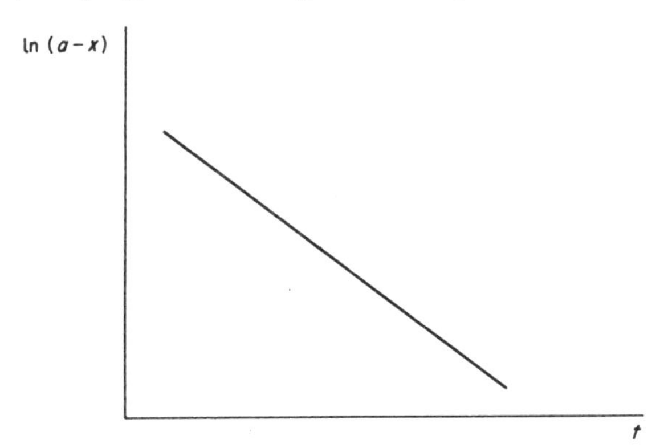

FIG. 4.1. Graph of $\ln(a-x)$ against t for a first order reaction.

reaction. The slope of the line is $-k$, the velocity constant of the reaction (Fig. 4.1).

In a reaction of *second order*, for example

$$A+B=X+Y,$$

the reaction rate can be expressed as

$$-\frac{dc_A}{dt}=-\frac{dc_B}{dt}=k \cdot c_A \cdot c_B, \quad (4.4)$$

where c_A and c_B are the concentrations of A and B respectively at time t. The rate at which c_A diminishes must be the same as the rate at which c_B diminishes in such a reaction.

Alternatively, if a moles of A and b moles of B are present initially in unit volume and x moles have reacted in unit volume in time t, (4.4) can be written

$$-\frac{d(a-x)}{dt}=-\frac{d(b-x)}{dt}=k(a-x)(b-x),$$

or

$$\frac{dx}{dt}=k(a-x)(b-x). \quad (4.5)$$

If, originally, A and B are present in equimolar amounts, $a=b$ and

$$\frac{dx}{dt}=k(a-x)^2=k(b-x)^2. \quad (4.6)$$

It will be seen that, where the reactants are originally present in equimolar amounts, (4.6) can be generalized as

$$\frac{dx}{dt}=k(a-x)^n \quad (4.7)$$

where $n =$ the order of the reaction, which can be determined experimentally, and can be zero, an integer or a fraction, depending on the mechanism of the reaction. Integration of (4.7) gives (except for a value of $n = 1$)

$$\frac{(a-x)^{1-n}}{n-1} = kt + \text{constant}, \tag{4.8}$$

and a graph of $(a-x)^{1-n}$ against t should give a straight line from whose slope k can be determined.

4.4. Determination of the Order and Velocity Constant of a Reaction

The concentrations of reactants or products are measured at various times during the progress of a reaction and several methods of calculating the order and velocity constant can be used. The first is the *Method of Integration*. The integrated forms of the rate equations, such as (4.3) and the general equation (4.8), are used to see whether one of them gives a value of k which does not vary as t increases. This "trial and error" method is not very sensitive in distinguishing between orders, especially where a complex reaction is being studied or measurements have not been made over a large proportion of the complete reaction time. Fractional orders or changing orders tend to be missed.

The *Half-Life Method* determines $t_{1/2}$, the time taken for half the original reactants to be used up. In a first order reaction, using (4.3), $x = a/2$ and

$$kt_{1/2} = \ln \frac{a}{a-a/2},$$

so that

$$k = \frac{1}{t_{1/2}} \ln 2. \tag{4.9}$$

Notice that for a first order reaction the half-life is independent of the initial concentration of the reactants, an important characteristic. For a second order reaction,

$$k = \frac{1}{at_{1/2}}, \tag{4.10}$$

and for a reaction of order n, using (4.8),

$$k = \frac{2^{(n-1)} - 1}{t_{1/2}(n-1)a^{(n-1)}}. \tag{4.11}$$

These relationships are obeyed provided that the reactants are initially at the same concentration, a, and if we look again at (4.11), we see that $t_{1/2} \propto \frac{1}{a^{(n-1)}}$. Therefore, using two initial concentrations a_1 and a_2 in separate experiments,

$$\frac{{}_1t_{1/2}}{{}_2t_{1/2}} = \frac{a_2^{(n-1)}}{a_1^{(n-1)}},$$

and, taking logarithms and rearranging terms,

$$n = 1 + \frac{\log({}_1t_{1/2}/{}_2t_{1/2})}{\log(a_2/a_1)}. \tag{4.12}$$

This method could be applied to any other convenient fraction of the total reaction than the half-life.

The *Differential Method* depends on (4.1) being generalized for a reaction of order n involving only one reacting substance,

$$-\frac{\mathrm{d}c}{\mathrm{d}t} = kc^n. \tag{4.13}$$

For two different initial concentrations c_1 and c_2,

$$-\frac{\mathrm{d}c_1}{\mathrm{d}t} = kc_1^n, \qquad -\frac{\mathrm{d}c_2}{\mathrm{d}t} = kc_2^n.$$

Taking logarithms and subtracting the two expressions,

$$\log\left(-\frac{\mathrm{d}c_1}{\mathrm{d}t}\right) - \log\left(-\frac{\mathrm{d}c_2}{\mathrm{d}t}\right) = n(\log c_1 - \log c_2). \qquad (4.14)$$

Thus n can be determined by taking two values of concentration and measuring the rate of reaction at these concentrations. If more than one substance is involved in the reaction, experiments are carried out in which only one reactant concentration varies. The order of reaction with respect to this reactant is determined, and added to the order with respect to the other reactant determined in a similar manner.

An extension of this method is to have all the reactants except one in excess, so that the rate is dependent only on the one reactant. In this way the order of reaction with respect to each reactant can be determined separately. The results are unreliable because with complex reactions, different mechanisms may come into play under the different conditions of "isolation" of the various reactants.

Reference 2 should be consulted for further details of experimental methods.

4.5. Reversible Reactions

We have so far considered reactions which progress only in one direction—and where a system is far from equilibrium this approach is justified—but when a system is close to equilibrium, the rate of the reverse reaction must be considered in determining the overall rate of reaction.

In the hypothetical first order reaction

$$A \rightleftharpoons B,$$

the rate of the forward reaction is, according to (4.7),

$$\frac{\mathrm{d}x}{\mathrm{d}t} = k_1(a-x) \qquad (n=1)$$

and the rate of the reverse reaction is

$$\frac{dx}{dt} = -k_2x,$$

where k_1 and k_2 are the velocity constants of the forward and reverse reactions respectively, and $(a-x)$ and x are the concentrations of A and B respectively at time t.

The overall rate of reaction in the forward direction is therefore

$$\frac{dx}{dt} = k_1(a-x) - k_2x. \tag{4.15}$$

At equilibrium, the overall rate of reaction is zero, and

$$k_1(a-x') = k_2x', \tag{4.16}$$

where x' is the net amount of A which has reacted to form B at the position of equilibrium. Therefore

$$\frac{k_1}{k_2} = \frac{x'}{a-x'} = K, \tag{4.17}$$

the equilibrium constant of the reaction.

Substituting the value of k_2 obtained from (4.16) in (4.15),

$$\frac{dx}{dt} = k_1(a-x) - \frac{k_1x(a-x')}{x'}$$

$$= \frac{k_1a}{x'}(x'-x). \tag{4.18}$$

Integrating,

$$\frac{k_1a}{x'} = \frac{1}{t}\ln\frac{x'}{x'-x} \tag{4.19}$$

which allows k_1 to be determined by measurement of x' and of the variation of x with t. k_2 can then be determined using (4.16).

This treatment of reversible reactions close to equilibrium has been extended by L. S. Darken[3] to include the two-stage reaction by which carbon is removed from liquid iron in the open hearth furnace.

4.6. The Effect of Temperature on Rates of Reaction

There is a significant relationship between the rate of a chemical reaction and the temperature at which the reaction takes place. For homogeneous reactions, it is usually found that the velocity constant of the reaction is approximately doubled for every ten degree rise in temperature, and the relationship can be represented by the equation

$$k = A \exp\left(-\frac{E}{RT}\right), \tag{4.20}$$

which is the *Arrhenius Equation* (proposed by S. Arrhenius in 1889), where k is the velocity constant, $\boldsymbol{R}$ the gas constant, E the *activation energy* of the reaction, and A is known as the *frequency factor*. The reasons for the names for E and A will become apparent later in this chapter.

The Arrhenius equation can be applied successfully to many homogeneous gas reactions, to reactions in solution and to heterogeneous reactions, and can be used in several forms such as

$$\ln k = \ln A - \frac{E}{RT}. \tag{4.21}$$

A graph plotted of ln k against $1/T$ gives a straight line from whose slope E can be determined.

Differentiating (4.21) with respect to temperature gives

$$\frac{\mathrm{d} \ln k}{\mathrm{d} T} = \frac{E}{RT^2}, \tag{4.22}$$

as A is a constant, and the greater the activation energy of a reaction, the greater will be the increase of reaction rate with temperature.

The relationship between the equilibrium constant and temperature in the van't Hoff isochore (2.40) has a form similar to (4.22), and it was van't Hoff who first argued (1887) that the logarithm of the velocity constant should bear a linear relationship to the reciprocal of the absolute temperature (4.21). Arrhenius extended this argument and interpreted (4.20) by suggesting that reactant molecules must be "activated" by acquiring an extra energy E, the activation energy of the reaction, before they could react. This interpretation is still accepted and will be developed in Sections 4.7, 4.8 and 4.9, where the important theories of reaction kinetics are outlined.

Some examples of the magnitude of the activation energy E in metallurgical reactions will indicate the important control this property of a reaction has on the reaction rate. A very simple homogeneous gas reaction, the thermal decomposition of hydrogen iodide

$$2HI = H_2 + I_2$$

has an activation energy of 184 kJ/mol.[2] This value is similar to that obtained for many phase changes and other phenomena in solid metals and alloys. Taking a value of 180 kJ, from (4.20), the rate at room temperature (300°K) will be proportional to exp $(-E/\boldsymbol{R}T)$, so that, taking $\boldsymbol{R}$ as 8·314 J/°K mol,

$$\text{rate}_{300} \propto \exp\left(-\frac{180{,}000}{2494}\right).$$

If the temperature was raised by 100°C, the rate would then be

$$\text{rate}_{400} \propto \exp\left(-\frac{180{,}000}{3326}\right).$$

This means that raising the temperature by 100°C will raise the reaction rate by a factor of

$$\frac{\text{rate}_{400}}{\text{rate}_{300}}=\exp\left(-\frac{180{,}000}{3326}+\frac{180{,}000}{2494}\right)=6{\cdot}9\times10^{7}.$$

This explains why transformations in solid alloys which do not take place at room temperature can be induced to take place if the temperature is raised slightly (precipitation hardening, tempering of martensite, recrystallization, etc.).

If the activation energy of the process mentioned above was halved, to 90 kJ/mol, the effect of a 100°C rise in temperature from room temperature would be to raise the reaction rate by a factor of about 8×10^3. This emphasizes the important control the activation energy has on the temperature-dependence of the reaction rate.

Kurdjumov[4] quotes figures showing that the half-life of the first stage in the tempering of martensite in a 1%C steel is 340 years at 0°C and 45 sec. at 160°C. This indicates an activation energy of 119·7 kJ/mol for the process. The activation energy for the diffusion of carbon in α-iron is 84·1 kJ/mol,[5] and that for the "self-diffusion" of α-iron (the diffusion of the iron atoms themselves) is approximately 280 kJ/mol,[5] so that it is a possibility that an important rate-controlling factor in the tempering process could be the diffusion of carbon in the iron to cause a rearrangement of the structure, and probably not the diffusion of iron atoms. This is an oversimplification of a complex phenomenon, but is an example of the diagnostic possibilities of measurements of activation energies of processes.

Slag/metal reactions in iron- and steelmaking often have activation energies of the order of 105 kJ/mol, but the value of 544 kJ/mol[6] for the transfer of silicon from blast-furnace type slags to liquid iron is very high. This may be due to the difficulty in breaking the strong covalent silicon–oxygen bonds holding the silicon atoms in the slag structure. Ward[7] draws attention to the similarity between this value for the activation energy and the energy of a silicon–oxygen bond.

Austenite, the interstitial solid solution of carbon in face-centred cubic γ-iron, will transform to a structure known as bainite if quenched to a suitable temperature and then allowed to transform isothermally. Bainite consists of ferrite, the interstitial solid solution of carbon in body-centred α-iron, and iron carbide. Radcliffe and Rollason[8] studied the kinetics of this transformation in the range 200–450°C by measuring the changes in electrical resistivity of iron wires of various carbon content quenched from 1100°C to various temperatures and allowed to transform isothermally at the temperatures to which they were quenched. By this means the progress of transformation could be followed.

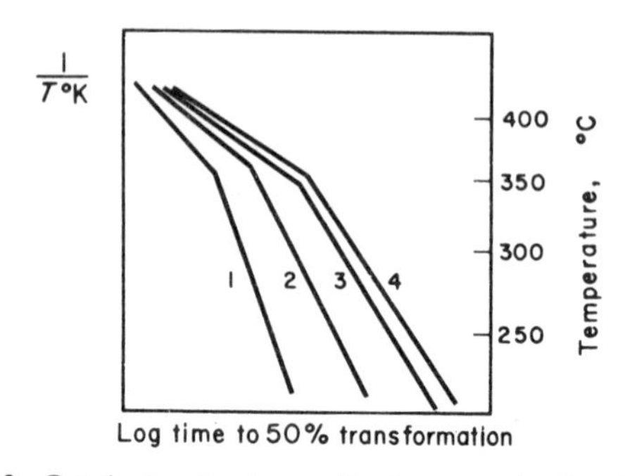

FIG. 4.2. Graph showing log $t_{1/2}$ for the austenite–bainite transformation against the reciprocal of the absolute temperature for a number of iron–carbon alloys and a plain carbon steel. 1. Plain carbon steel, 0·65%C. 2. Iron–carbon alloy, 0·77%C. 3. Iron–carbon alloy, 1·07%C. 4. Iron–carbon alloy, 1·20%C. (After Radcliffe and Rollason.[8])

Using the Arrhenius equation in the form of (4.21),

$$\ln k = \ln A - \frac{E}{RT},$$

a graph was plotted of the logarithm of the time taken for 50% of the material to transform to bainite, which is the half-life and is a measure of the rate of transformation, against $1/T$ for the different steels. It can be seen from Fig. 4.2 that the results fall on two straight lines intersecting in the region of 350°C for each material examined, and it was concluded that there are marked differences between the mechanisms of bainite transformations above and below about 350°C. The values for the activation energy E for the transformations can be obtained from the slopes of these curves, and the value is about twice as high for the transformation above 350°C as that below 350°C. For example, for a 0·65%C steel, E is 77·4 kJ/mol above 350°C and 31·4 kJ/mol below 350°C. These results support the conclusions from electron-microscope observations of bainite structures—that there are two forms of bainite, "upper" and "lower" bainite, depending upon the temperature at which the bainite is formed by the isothermal transformation of the austenite.

The further implications of this work are beyond the scope of the present book, but this last example has been used to illustrate the method by which the measurement of activation energies can be used to study transformations taking place in the solid state, and to back up conclusions drawn from the microscopic examination of the transformation products. It also illustrates the possibilities offered by this approach for detecting and perhaps even identifying variations in reaction mechanism.

4.7. Theories of Reaction Kinetics: The Collision Theory

A perfect theory of chemical kinetics should provide a method of calculating the rate of any reaction from certain basic principles. Such a theory has not been produced, and there are many difficulties standing in the way of the final calculations in addition to the complexity of elucidating a basic theory. However, we are able to understand certain factors which affect reaction rates and can sometimes predict their values with the aid of existing theories.

Considering a *homogeneous* bimolecular reaction (in the gas phase) between two hypothetical molecules A and B, it is reasonable to suppose that, for reaction to occur, the molecules A and B must collide. The *Collision Theory* of reaction rates (due mainly to work of W. C. McC. Lewis and C. N. Hinshelwood about 1920) uses the kinetic theory of gases to calculate the number of times one molecule of A collides with one of B. We will call the total number of collisions per second taking place between A and B in one cubic centimetre Z.[9] Then, if we assume that each time a molecule of A collides with one of B the reaction occurs, it is possible to calculate the velocity constant of the reaction. Such a calculation can indicate a rate of reaction as much as 10^{17} times that actually measured – indicating that only one collision in 10^{17} results in reaction. When the effect of temperature on reaction rate is considered, the number of collisions taking place per second is proportional to the square root of the absolute temperature according to the kinetic theory of gases, indicating an increase of about 1% in the collision frequency for a ten degree rise in temperature. In fact, a ten degree rise in temperature frequently doubles the rate of reaction. It will be clear that the assumption that every collision results in reaction is invalid, and it is suggested that, for reaction to occur, A and B must possess a certain minimum relative translational energy when they collide.

Referring to the Arrhenius equation (4.20), it can be considered that this minimum relative translational energy requirement is E, the activation energy of the reaction. Because collisions between molecules in a gas are continually occurring, involving a transfer of energy from one to another, there will be a Maxwellian distribution of molecular energies (Fig. 4.3). Then there will be a certain fraction of molecules present possessing sufficient energy (E) to react. The shape of this distribution curve changes with temperature, and the effect of an increase of ten degrees on the fraction of collisions of molecules possessing at least relative energy E can be calculated.

Although the total number of collisions only increases by about 10%, the fraction of collisions of molecules possessing at least relative energy E is greatly increased – thus explaining the discrepancy mentioned above. This relationship between the fraction of collisions of molecules possessing a minimum relative energy E in a distribution such as that shown in Fig. 4.3 and the temperature is given by the Boltzmann factor, $\exp(-E/\boldsymbol{R}T)$, so that the actual number of collisions resulting in reaction per second in one cubic centimetre, which is a measure of the reaction rate, is $Z \exp(-E/\boldsymbol{R}T)$, and the relationship between this expression and the Arrhenius equation (4.20) is established.

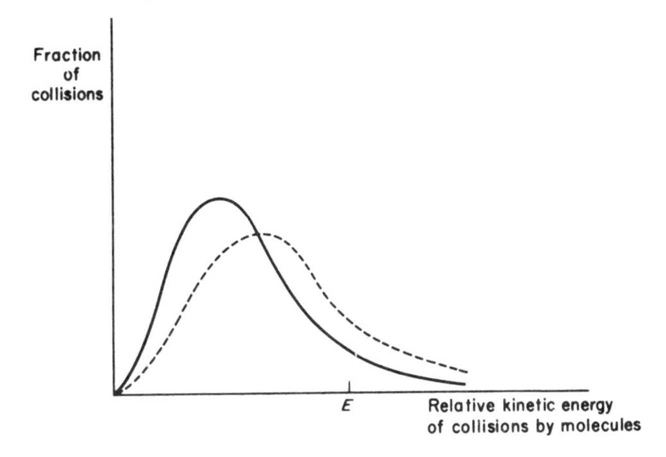

FIG. 4.3. Maxwellian distribution of the energies of molecular collisions. Dotted curve shows the effect of a rise in temperature on the distribution. The area beneath the curves to the right of E represents the fraction of collisions possessing at least energy E.

Although reaction rates for a few bimolecular reactions involving simple molecules calculated by the application of collision theory correspond to the observed reaction rates, in many cases the calculation is wildly inaccurate – involving a factor of as much as 10^5 times the experimental results.

The reason for the failure of the collision theory to predict reaction rates in all but a few simple cases is the basic assumption that the colliding molecules behave as rigid spheres, and that the collisions involve only changes in kinetic energy. In simple reactions this may be approximately true, but where complex molecules are involved the rotational and vibrational energy and the orientation of the reacting molecules must be taken into account. It has not been possible to calculate the fraction of collisions resulting in reaction because of the complexity of the calculations involved, and consequently the collision theory has been replaced by other theories which have had greater success in predicting reaction rates. Nevertheless, the pictorial simplicity of the collision theory has great attractions and serves as a useful introduction to the problems involved in the study of reaction kinetics. An attempt has been made to account for the complexity of a collision by introducing a "steric probability factor" P (which can vary from 1 to 10^{-8}) by which the collision number Z must be multiplied to give the correct magnitude of reaction rate. The steric probability factor cannot be calculated from basic principles, and seems to be an unsatisfactory compromise. Hinshelwood[10] can be consulted for a detailed treatment of the collision theory of reaction rates.

4.8. The Activated Complex

The work of Eyring, Polanyi, London and others in the 1920's and 1930's resulted in a more successful approach to reaction kinetics.[11]

Taking as an example the bimolecular reaction between hydrogen and iodine to form hydrogen iodide,

$$H_2 + I_2 = HI + HI,$$

if a molecule of hydrogen approaches a molecule of iodine, the nuclei of the hydrogen and iodine atoms will be forced apart because there will be some attraction between the atoms of the two molecules. This produces an increase in potential energy of the two molecules at the expense of kinetic energy. The molecules, if they do not possess sufficient energy to react, will then fly apart. A graph can be plotted of the energy of the system containing the two molecules against the distance travelled or the time—either of which can be called the "reaction coordinate" (Fig. 4.4).

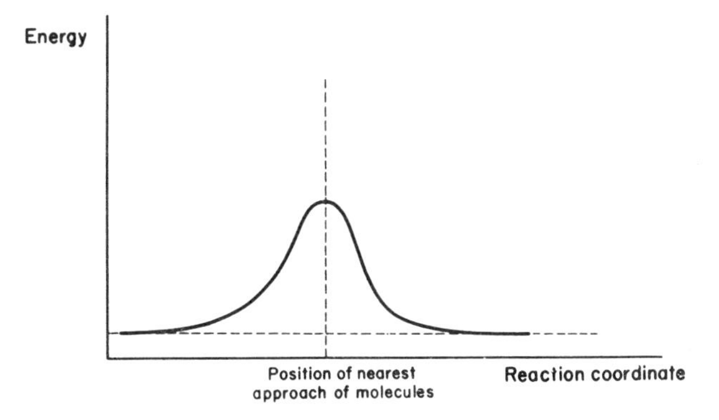

FIG. 4.4. Energy change during collision without reaction.

If the molecules of hydrogen and iodine possess sufficient energy to react, they will approach one another so closely that, provided they are also correctly orientated, they will react to form an *activated complex,* which can be regarded as a molecule having only a temporary existence, and which then decomposes at a definite rate to form the products of the reaction (Fig. 4.5).

The activated complex will be in a high energy state, as it possesses all the energy that the reacting molecules had before the collision, and the energy changes in the reaction processes can be expressed in a diagram such as Fig. 4.6. As the total energy of the products is greater than that of the reactants (ΔH positive), this is an endothermic reaction, and the difference between the energy of the reactants and the energy of the activated complex is E, the activation energy of the reaction (see Section 4.6). The activation energy of the reverse reaction

$$2HI = H_2 + I_2$$

would be E^R, and the reaction is exothermic (ΔH negative) by the same process of reasoning.

The nature of the activated complex depends on the reacting substances, but it is formed from the reactants. The atoms in

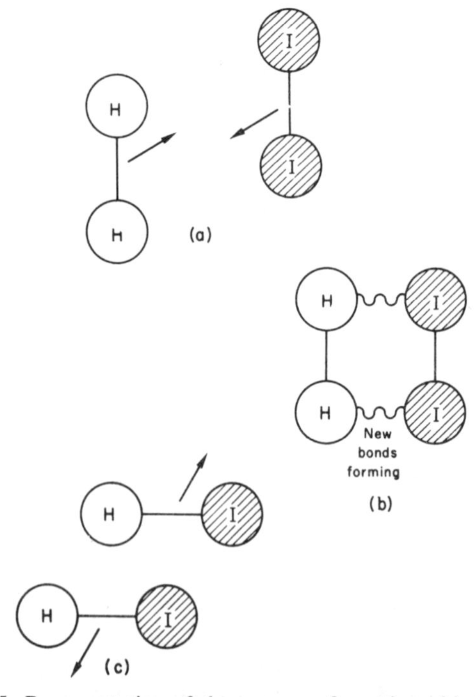

FIG. 4.5. Representation of the process of reaction of iodine and hydrogen to form hydrogen iodide. (a) Molecules of hydrogen and iodine approach one another in correct orientation and with sufficient energy to react. (b) Activated complex formed. (c) Products of reaction (hydrogen iodide molecules) fly apart as activated complex decomposes.

the activated complex are generally less densely packed than they are in the reactant molecules and their motion is more disordered. The frequency of oscillation of the atoms is lower, the strength of the interatomic bonds is lower and they are different from those in the reactants. The energy of the activated complex is higher so that the energy of the system increases when the complex forms, but not sufficiently to completely break the interatomic bonds involved. The complex undergoes translational, oscillatory and rotational motion

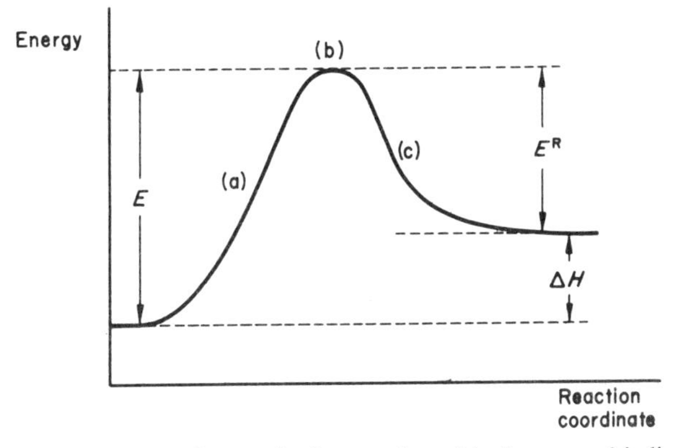

FIG. 4.6. Energy changes in the reaction of hydrogen and iodine to form hydrogen iodide. (a), (b) and (c) refer to the stages shown in Fig. 4.5.

which can, under certain circumstances, lead to its breakdown —either into the original reactant molecules or into the products of the reaction. The whole process can be regarded as a redistribution of energy among the participating species, and an equilibrium is set up between the reactants and the activated complex so that the thermodynamic properties of the equilibrium can be postulated. In many processes the name "activated complex" is perhaps difficult to explain, and the alternative *transition state* is a more accurate description.

4.9. The Theory of Absolute Reaction Rates

Having established the concept of the activated complex, Erying and others then used it to calculate the rate of a chemical reaction from the frequency at which molecules, possessing sufficient energy and having the correct orientation, form the activated complex for the reaction, which then decomposes at a definite rate. This approach is known as *the theory of absolute reaction rates,* but it has also been called *the transition state theory* of chemical kinetics, or simply *the activated complex theory.* This theory was so called because it was originally considered that reaction rates could be calculated absolutely, knowing only the basic properties of the reactants. The formation of an activated complex can be considered to occur in all chemical reactions, and in certain physical processes (for example, diffusion in solid metals, which will be discussed later), but for the moment we will continue to consider the case of the simple bimolecular gas reaction.

It is possible to derive an equation for the velocity constant k of the overall reaction in terms of K^*, the equilibrium constant of the reaction between the reactants to form the activated complex (Eyring).

$$k = \frac{\boldsymbol{RT}}{\boldsymbol{Nh}} K^*, \tag{4.23}$$

where $\boldsymbol{R}$ is the gas constant, T the absolute temperature, $\boldsymbol{N}$ Avogadro's number, and $\boldsymbol{h}$ is Planck's constant. (Note that we will use the asterisk * to denote a symbol applying to the reaction by which the activated complex is formed,

$$\text{reactants} \overset{*}{\rightleftharpoons} \text{activated complex.})$$

$\boldsymbol{RT}/\boldsymbol{Nh}$ is a universal factor, varying only with the temperature and bearing no relationship to the substances reacting or to the activated complex involved in the reaction. The equilibrium between the reactants and the activated complex is not much disturbed by the relatively small amount of decomposition of the activated complex to form the products of the reaction.

Using the expressions

$$\Delta G^{\circ} = -RT \ln K_c \tag{2.39}$$

and

$$\Delta G^{\circ} = \Delta H^{\circ} - T\Delta S^{\circ}, \tag{2.18}$$

referring ΔG° to a standard state of concentration rather than pressure, we can write

$$\Delta G^* = -RT \ln K^* \tag{4.24}$$

and

$$\Delta G^* = \Delta H^* - T\Delta S^*, \tag{4.25}$$

where ΔG^*, ΔH^* and ΔS^* are understood to be the *standard* free energy, enthalpy and entropy changes respectively involved in the formation of the activated complex – omitting the superscript $^{\circ}$ in each case.

Then, from (4.24),

$$K^* = \exp\left(-\frac{\Delta G^*}{RT}\right),$$

and hence, using (4.23),

$$k = \frac{RT}{Nh} \exp\left(-\frac{\Delta G^*}{RT}\right). \tag{4.26}$$

The value of the free energy of activation, ΔG^*, can be replaced in (4.26) so that

$$k = \frac{RT}{Nh} \exp\left(-\frac{\Delta H^* - T\Delta S^*}{RT}\right)$$

$$= \frac{RT}{Nh} \exp\left(\frac{\Delta S^*}{R}\right) . \exp\left(\frac{-\Delta H^*}{RT}\right), \tag{4.27}$$

and ΔH^*, the enthalpy of formation of the activated complex, corresponds to E, the activation energy of the Arrhenius equation (4.20). The frequency factor, A, in the Arrhenius equation can then correspond to the term $(\boldsymbol{RT}/\boldsymbol{Nh}) \,.\, \exp(\Delta S^*/\boldsymbol{R})$, and $\exp(\Delta S^*/\boldsymbol{R})$ plays the same part as the steric probability factor, P used to explain deviations from the Simple Collision Theory (Section 4.7).

Where simple molecules are involved, ΔS^*, the entropy change (measuring the change in the state of ordering in the system) when the activated complex is formed, will be very small and $\exp(\Delta S^*/\boldsymbol{R})$ will be approximately unity. Where complex reactant molecules are involved, there will probably be a large decrease in entropy and $\exp(\Delta S^*/\boldsymbol{R})$ will be very small.

Summarizing, we can write down three equations relating the velocity constant of a reaction to the absolute temperature:

Arrhenius' equation, $$k = A \exp\left(-\frac{E}{\boldsymbol{R}T}\right), \qquad (4.20)$$

Collision Theory equation, $$k = ZP \exp\left(-\frac{E}{\boldsymbol{R}T}\right),$$

Theory of Absolute Reaction

Rates equation, $$k = \frac{\boldsymbol{R}T}{\boldsymbol{N}\boldsymbol{h}} \exp\left(\frac{\Delta S^*}{\boldsymbol{R}}\right) \,.\, \exp\left(-\frac{\Delta H^*}{\boldsymbol{R}T}\right). \qquad (4.27)$$

A comparison between these equations should demonstrate the relationship between the three approaches.

4.10. Unimolecular Reactions

In a reaction such as

$$AB = A + B,$$

the theory of absolute reaction rates still holds, but it is necessary to consider how the molecule AB attains the activated state from which it can decompose to A and B. It has been suggested that although the AB molecules become activated by energy transfer in collisions with other molecules (as in the discussion in Section 4.7), there may be a delay before decomposition begins. In this period, the activated AB^* molecule may come into collision with another molecule – losing its extra energy once again and becoming deactivated before it can decompose. This accounts for the very high activation energies associated with many unimolecular reactions. The theory of unimolecular reactions will not be discussed further in this volume, but several books can be consulted.[12] The titles under ref. 12 can be used for further reading on the theories of reaction kinetics, and are acknowledged as the source of much of the information to this stage in this chapter.

4.11. Catalysis

When the rate of a reaction is changed by the presence of a substance, which is not itself altered chemically as a result of the reaction, this substance is known as a *catalyst*. A substance which causes an acceleration of the reaction is usually known simply as a catalyst, whereas a "negative catalyst" reduces the reaction rate. It should be noted that although the catalyst is unchanged chemically, it may undergo a physical change – such as the reduction from a coarse crystalline state to a fine powder. If the reaction is reversible, a catalyst will affect the reverse reaction to the same extent as the forward reaction, and will not change the equilibrium constant of the reaction. Although large increases in reaction rates can be brought about by comparatively small quantities of catalyst, the rate of a catalysed reaction is in general proportional to the quantity of catalyst present. This is extended in the case of gas reactions at the surface of solid catalysts to include the surface area of the catalyst – the greater the area in general, the greater the catalytic effect.

It is considered that catalysts actually take a part in the reaction – being involved in the formation of the activated complex in the appropriate stages of the reaction – and are then released chemically unchanged when the reaction is completed. This applies whether the catalytic process is *homogeneous* (the catalyst being of the same phase as the reactants) or *heterogeneous* (the catalyst being of a different phase from that of the reactants). An important example of homogeneous catalysis is *acid-base* catalysis in the liquid phase. Hydrogen and hydroxyl ions are known to accelerate certain reactions in solution, and therefore any substance which will produce hydrogen or hydroxyl ions in solution can act as a catalyst in this manner. The reaction of nitrogen and hydrogen to form ammonia at the surface of iron under pressure at 500°C (the Haber process for the industrial production of ammonia) is an example of heterogeneous catalysis, and we will consider heterogeneous catalysis of gas reactions at the surface of a solid in more detail.

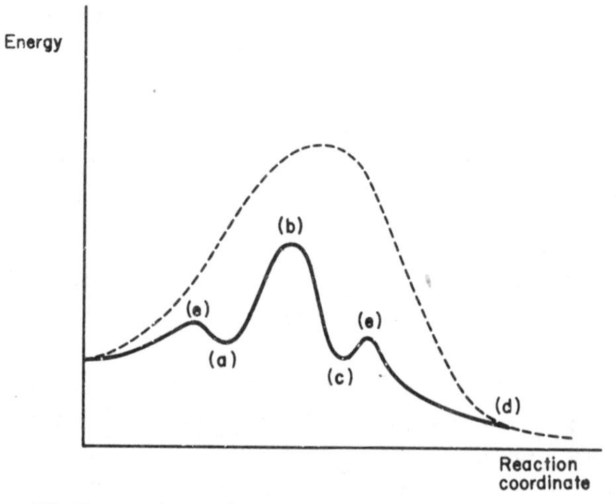

FIG. 4.7. Energy changes in a gas reaction. Broken curve represents the uncatalysed gas reaction, solid curve represents the gas reaction catalysed at a solid surface. (a), (b), (c), (d) and (e) are explained in text. (After Glasstone,[12] p. 1146.)

Referring to Fig. 4.7, it is considered that the effect of the catalyst is to reduce the overall activation energy of the reaction. In this case the reactants are adsorbed on the surface of the catalyst (a), and they then react—involving the catalyst itself in the reaction—to form an activated complex (b). The activated complex decomposes to form the products of the reaction (c)—still adsorbed on the surface of the catalyst, and the products are then desorbed (d). It will be seen that the overall activation energy in the catalysed reaction is less than the activation energy of the uncatalysed reaction. (e)–(e) are the energy humps through which the system passes when the reactants approach the surface (adsorption) or the products leave the surface (desorption). Clearly it is incorrect to say that a catalyst "plays no part in a reaction".

The surface of the catalyst is of importance in the reaction, because catalysts tend to be specific—one catalyst being more effective than another for a given reaction. There is no such thing as a universal catalyst which is equally effective for all gas reactions. Reaction does not take place wherever the reactants are adsorbed on the surface but only occurs at a small proportion of the positions of adsorption—the positions at which reaction does occur are called "active centres". These active centres could be at the peaks in an otherwise flat surface, at corners or edges of crystals, at grain boundaries or cracks, and it is interesting that the orientation of a metal surface has an important effect on the catalytic activity of that surface. It has been found that, although the extent of adsorption on the (110) planes of nickel is much the same as that on (100) and (111) planes, the catalytic effect of the (110) planes is much greater than that of the other planes—demonstrating that there must be more active centres on the (110) planes than on the others.

The reactants are adsorbed on the surface of the catalyst as a monolayer (a layer one molecule thick), and the forces involved are equivalent to chemical bonds.[13] We can thus consider that the catalytic surface is taking part in the reaction and that certain parts of this surface will possess sufficient energy to form the activated complex of the reaction with the reactants—these parts will be the "active centres". If one of the reactants or products is very strongly adsorbed, it can prevent the access of other reactants to the surface, and the reaction will slow down (see Section 4.2 for an example of such a situation—which leads to a zero order of reaction). Sometimes other substances, not involved in the reaction, can be strongly adsorbed and destroy the active centres—thus inhibiting the reaction. Such substances are known as "catalytic poisons", sulphides being particularly active poisons for metal catalysts. Small additions of certain substances to a catalyst can increase its catalytic activity—these substances are called "promoters". In the Haber process for the synthesis of ammonia, the iron catalyst is made much more effective by the addition of small amounts of potassium and aluminium oxides. This action of promoters has not been satisfactorily explained. The effect of alloying elements on the catalytic action of metals was studied by G. M. Schwab,[14] and his work indicated that the catalytic action of metals is associated with their electronic structures. As the number of "free" electrons increases with the addition of multivalent elements to a monovalent metal, the catalytic activity of the alloy decreases. This is an interesting approach to the study of metals, and links up with the electron theory of metals—a subject beyond the scope of this book.[15]

An example of the heterogeneous catalysis of a gas reaction may possibly occur in the blast furnace extraction of iron. Below about 900°C, there is excess carbon monoxide in the furnace gases, causing a reversal of the reaction

$$CO_2 + C = 2CO$$

and consequent deposition of carbon by the furnace gases. The reaction takes place slowly, but may be catalysed by iron or iron oxides (the definite identity of the catalyst is a subject of some controversy, but there is no doubt that iron is involved in one form or another – whether as oxide, carbide or elemental iron).[16] Deposition of carbon tends to occur preferentially at the surface of ore lumps or of refractory bricks containing iron, and can result in the disintegration of these materials where carbon monoxide has penetrated cracks and pores before decomposing. This disintegration can be the source of severe refractory wear and recent changes in the types of refractory used in blast furnace stacks have aimed at the removal of iron from the refractory mix – hence removing the possibility of the catalysis of the carbon deposition reaction within the brick itself.[17]

A more advanced treatment of catalysis by solid surfaces is given by Gray[18] in his discussion of defect structure and catalysis.

4.12. Diffusion

Gases, liquids and solids tend to intermix in the absence of mechanical agitation and convection currents, and when a reaction occurs in one place between two substances, the concentrations of those substances are lowered at the position of reaction. For the reaction to continue at that position, the reactants must be replenished by diffusion, mechanical agitation or convection.

Fick's laws (1855) form the basis of diffusion theory – applying to a two-component, single-phase solution at constant temperature and pressure. Fick's First Law can be expressed in the form

$$J = -D\frac{dc}{dx}, \qquad (4.28)$$

where J is the mass of diffusing substance passing in unit time through unit area of a plane at right angles to the direction of diffusion. J is known as the "flux", and is proportional to the concentration gradient dc/dx across the plane, where c is the concentration of the substance, x is the distance in the direction of diffusion (see Fig. 4.8). The constant of proportionality is D, the "diffusion coefficient" or "diffusivity", and is the value of J when unit concentration gradient exists at the plane of measurement. This law is similar to that relating to heat transfer by conductance and also the flow of an electric current along a conductor.

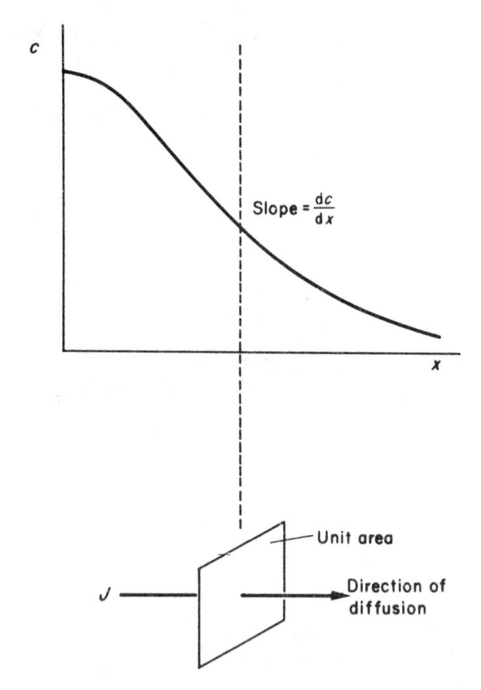

FIG. 4.8. Representation of Fick's First Law of diffusion.

Fick's Second Law of Diffusion can be expressed in the form

$$\frac{\partial c}{\partial t} = \frac{\partial}{\partial x}\left(D\frac{\partial c}{\partial x}\right), \tag{4.29}$$

where $\partial c/\partial t$ is the rate of increase in concentration of the diffusing substance in a given volume, and is equal to the difference between the flux entering the volume and that leaving the volume, $-\partial J/\partial x$.

If D is considered to be constant at constant temperature, (4.29) can be written

$$\frac{\partial c}{\partial t} = D\frac{\partial^2 c}{\partial x^2}, \tag{4.30}$$

but it is usually found that D is not constant and varies with the concentration. Since the measurement of flow rates under fixed concentration gradients can be difficult, it is more convenient to use the Second Law and measure the change in concentration with time. The differential equations produced can be solved under certain boundary conditions which can be approximated closely in experimental procedures.[19]

The diffusion coefficient D is related to temperature by an expression similar to Arrhenius' equation (4.20)

$$D = D_0 \exp\left(-\frac{E_D}{RT}\right), \tag{4.31}$$

where E_D is the activation energy for diffusion and can be represented in the solid state as the energy required to move the diffusing particle from its low energy position, force neighbouring particles apart, and pass through a transition state similar in significance to an activated complex (see Fig. 4.9).

With the close-packed structures of solid metals, E_D is large and D is strongly temperature-dependent. In the liquid state the atoms or ions are less tightly packed and E_D will be smaller – so that diffusivities in liquids do not vary as much as in solids over a wide range of temperatures. For example, the value of E_D for carbon in solid iron is of the order of 80 kJ/mol,[5] and in liquid iron it is 40–60 kJ/mol.[20] The activation energy for self-diffusion in liquid lead is 18·8 kJ/mol and in liquid silver 34·1 kJ/mol,[20] whereas the activation energies for self-diffusion in solid lead and silver are 116·7 and 192·0 kJ/mol respectively.[5]

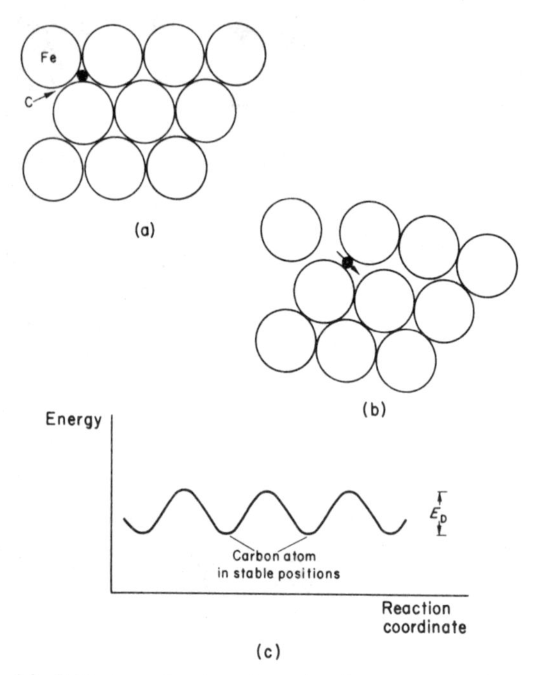

FIG. 4.9. Diffusion of carbon through a face-centred cubic lattice of γ-iron. (a) shows stable interstitial position of carbon atom in γ-iron lattice. (b) shows transition state when carbon atom diffuses—the neighbouring iron atoms are disturbed from their stable positions and this requires an increase in energy of the system. (c) graph representing change in energy of system as carbon atom diffuses through iron lattice. E_D is the activation energy of diffusion.

Fick's laws cannot apply directly where the medium through which a substance diffuses is not homogeneous—for instance, at a grain boundary in a metal, where diffusivity will vary with the orientation of the adjacent grains.

4.13. Diffusion in the Solid State[21]

Diffusion is rate-controlling in many metallurgical processes, for example the growth of pearlite from austenite in steels, oxide-film formation on metals, carburizing and the removal of inhomogeneous structures by annealing of cored single-phase alloys. The theory of the diffusion mechanism was built up originally from experimental work on ionic solids, but has subsequently been extended to cover metallic structures. In addition to the diffusion of a foreign particle in a substance, the "self-diffusion" of the substance within itself is of importance — this can be measured by the technique of using radioactive isotopes to "label" certain atoms and tracing their movement in the substance.

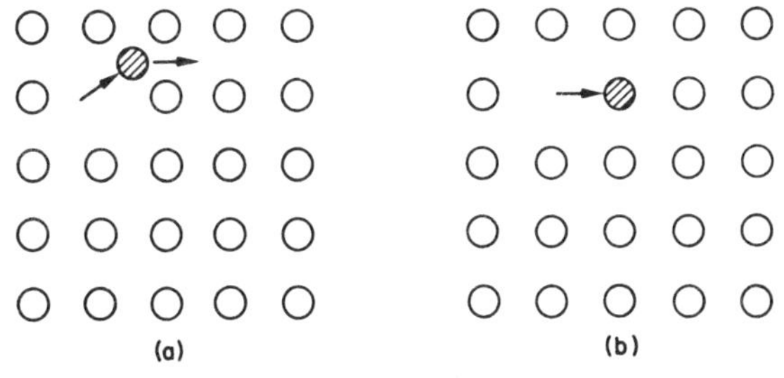

FIG. 4.10. (a) Frenkel defect diffusion. (b) Schottky defect diffusion.

The work on ionic solids led to the formulation of two mechanisms for solid state diffusion, the Frenkel defect (interstitial mechanism) (1926) and the Schottky defect (vacancy mechanism) (1930), (see Fig. 4.10). The diffusing atoms move through the interstitial spaces between the atoms in the normal lattice sites in diffusion by the Frenkel mechanism, and vacancy diffusion allows atoms to move around by receiving atoms from neighbouring sites into vacancies, and then following up by a similar movement by other neighbours into the vacant site.

A third mechanism requires exchange or rotation within a coplanar group of atoms forming a closed ring (Fig. 4.11).

A confirmation of the dependence of the creep of metals on self-diffusion, which causes recovery during the deformation process, was obtained by Dorn,[22] who found a good correlation between the activation energy for creep and the activation energy for self-diffusion for a number of elements, including

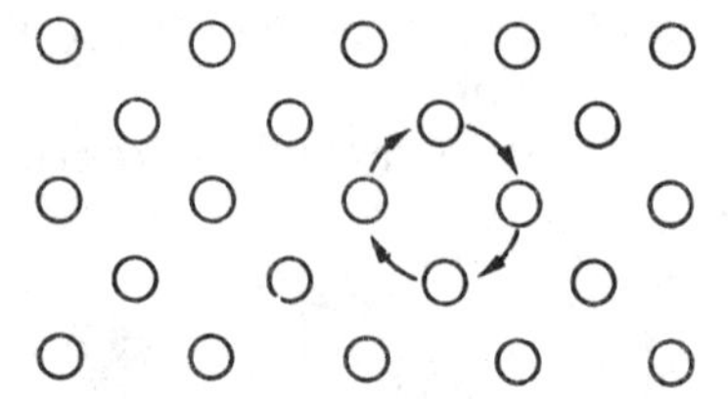

FIG. 4.11. Rotation of atoms in a closed coplanar ring in the (100) plane of a face-centred cubic lattice.

iron, gold, copper, aluminium, zinc, lead, cadmium and indium. Dislocations are supposed to climb away from slip planes to form sub-grain boundaries, and the mechanism by which they climb is thought to be vacancy diffusion.

4.14. Reaction Kinetics in Metallurgical Systems

To sum up this chapter, the main theoretical treatment has been that of the chemist, who considers reactions between reactant molecules and has developed theories which explain the behaviour of homogeneous reactions. In the last two sections of the chapter, it will have become apparent that ideas of "activated complexes" and simple gas reactions may not be so easily transferred to reactions in liquid and solid metals. Metallic structure is not one of "molecules" but of atoms or ions held in the structure by very mobile bonds. Ideas of collisions between reactants in the sense described in Section 4.7 may be difficult to hold in liquid and solid metallic systems, and as temperatures are often very high, the rates of the actual chemical reactions taking place will be correspondingly high. Only where extreme difficulty is experienced in the reaction step (for example, in the transfer of silicon between slag and liquid iron mentioned in Section 4.6) is the chemical reaction itself likely to be rate-controlling.

In order that a reaction may proceed, reactants and products must be brought to the reaction site and removed from the site, and we are therefore likely to find diffusion and other transport mechanisms rate-controlling—and this is especially the case in heterogeneous reactions where the reaction takes place at the interface between phases. Behaviour at such interfaces and transport mechanisms in heterogeneous reactions will be discussed in Chapter 6, which has been set aside because most metallurgical reactions are heterogeneous and a large fraction of metallurgy as a subject involves the study of the interface between phases.

References

1. WARD, R. G. *J. Iron and Steel Institute*, **201** (1963), 11.
2. LAIDLER, K. J. *Chemical Kinetics,* McGraw-Hill, New York, 1950, Chapters 1 and 2.
3. DARKEN, L. S. and GURRY, R. W. *Physical Chemistry of Metals,* McGraw-Hill, New York, 1953, p. 479.
4. KURDJUMOV, G. V. *J. Iron and Steel Institute*, **195** (1960), 41.
5. SMITHELLS, C. J. *Metals Reference Book, Vol. II*, 3rd edn., 1962, pp. 589–95.
6. FULTON, J. C. and CHIPMAN, J. *Trans. A.I.M.M.E.* **215** (1959), 888.
7. WARD, R. G. *An Introduction to the Physical Chemistry of Iron and Steel Making*, Edward Arnold, London, 1962, p. 173.
8. RADCLIFFE, S. V. and ROLLASON, E. C. *J. Iron and Steel Institute*, **191** (1959), 63.
9. The reader can find a derivation of the value of Z, the "collision number", in sections on the kinetic theory of gases in textbooks of physical chemistry such as GLASSTONE, S. and LEWIS, D. *Elements of Physical Chemistry*, Macmillan, London, 1960, p. 30.
10. HINSHELWOOD, C. N. *Kinetics of Chemical Change*, Oxford University Press, 1940.
11. GLASSTONE, S., LAIDLER, K. J. and EYRING, H. *The Theory of Rate Processes,* McGraw-Hill, New York, 1941.
12. GLASSTONE, S. *Textbook of Physical Chemistry*, Macmillan, London, 1953, p. 1107. TROTMAN-DICKENSON, A. F. *Gas Kinetics,* Butterworths, London, 1955, p. 45: ref. 2, p. 76: EYRING, H. and EYRING, E. M. *Modern Chemical Kinetics*, Chapman & Hall, London, 1965, p. 51.
13. This approach is following the work of I. LANGMUIR (1916), which forms the basis of all subsequent theory of the mechanism of heterogeneous catalysis.
14. SCHWAB, G. M. *Disc. Faraday Soc.* (Heterogeneous Catalysis), **8** (1950), 166–71.
15. HUME-ROTHERY, W. and RAYNOR, G. V. *The Structure of Metals and Alloys*, Institute of Metals, London, 1954, Part I, p. 1.
16. KLEMANTASKI, S. *J. Iron and Steel Institute*, **171** (1952), 176: TAYLOR, J. *J. Iron and Steel Institute*, **184** (1956), 1: DAVIES, W. R. *et al. Trans. Br. Ceram. Soc.* **56** (1957), 107.

17. ALDRED, F. H. and HINCHLIFFE, N. W. *J. Iron and Steel Institute*, **199** (1961), 243.
18. GRAY, T. J. *The Defect Solid State*, Interscience, New York, 1957, pp. 239–92.
19. See ref. 3, Chapter 18.
20. LING YANG and DERGE, G. *Physical Chemistry of Process Metallurgy*, Part I (ed. G. R. ST. PIERRE), Interscience, New York, 1961, p. 503.
21. BIRCHENALL, C. E. *Metallurgical Reviews of the Institute of Metals*, London, **3**, No. 11 (1958), p. 235: OSIPOV, K. A. *Activation Processes in Solid Metals and Alloys*, Edward Arnold, London, 1964.
22. DORN, J. E. Paper No. 9. N.P.L. Symposium on Creep and Fracture of Metals at High Temperatures, 1954.

CHAPTER 5

Electrochemistry

5.1. Introduction

In an electronic conductor, an electric current is the result of a net movement of electrons in the structure of the conductor when an electrical potential is applied. Electrons have negligible mass compared with the remainder of the structure, and the flow of electricity is not accompanied by a significant movement of matter. Electrolytic conductors contain mobile ions, which carry an electrical charge. When an electrical potential is applied, these ions will move in the direction appropriate to their charge. As in electronic conductance, the electric current is a movement of electrical charge, but this time the charge is carried by ions of significant mass and electrolytic conductance is accompanied by mass transfer. Conductors may exhibit a combination of electronic and electrolytic conduction, but frequently the proportion of one type is so dominating that it is permissible to assume that the conduction is entirely by the dominating mechanism.

This property of electrolytic conduction plays a major role in some of the processes and phenomena important to the metallurgist—corrosion and oxidation of metals, electroplating, electropolishing, electrolytic extraction and refining of metals, for example. We will first consider the nature of electrolytes and electrolytic conductance, and then their function in galvanic cells and electrolysis. Some metallurgical applications will be considered in more detail in later chapters (7 and 8).

5.2. Electrolytes

The most detailed study of electrolytes (an alternative name for electrolytic conductors) has been carried out on aqueous solutions, but fused salts and solids can also be of considerable importance as electrolytic conductors.

If we consider a crystalline solid, the salt MX, whose bonding is ionic (electrovalent) in character, the structure will consist of a lattice of ions M^+ and X^- (Fig. 5.1). This structure

$$
\begin{array}{cccccccc}
M^+ & X^- & M^+ & X^- & M^+ & X^- & M^+ & X^- \\
X^- & M^+ & X^- & M^+ & X^- & M^+ & X^- & M^+ \\
M^+ & X^- & M^+ & X^- & M^+ & X^- & M^+ & X^- \\
X^- & M^+ & X^- & M^+ & X^- & M^+ & X^- & M^+
\end{array}
$$

FIG. 5.1. Two-dimensional representation of the crystal lattice of the ionic salt MX. M^+ and X^- are positively and negatively charged ions of the elements M and X respectively. (Perfect or ideal structure.)

is stable because the attractive forces between the charged ions are exactly balanced by the repulsive forces due to the interaction between the electron fields of the ions. The repulsive forces can be represented by the mechanical analogue of two rubber balls in collision—the balls become momentarily stationary in the act of collision when the kinetic energy of their motion towards one another is balanced by the elastic energy due to the distortion of the structure of the rubber. If the internal energy U of the system containing the two ions of opposite charges is related to their distance apart r, a curve of the type shown in Fig. 5.2 will be obtained. This shows that the internal energy of the crystal reaches a minimum U_0 when the ions are a distance r_0 apart, at which the attractive forces are exactly balanced by the repulsive forces between the ions. U_0 is known as the "lattice energy" of the crystal in its stable state of minimum energy, and usually refers to 1 mole of the crystal.

Water is a "polar" solvent—the great affinity of oxygen for electrons induces a slight positive charge δ^+ on the hydrogen

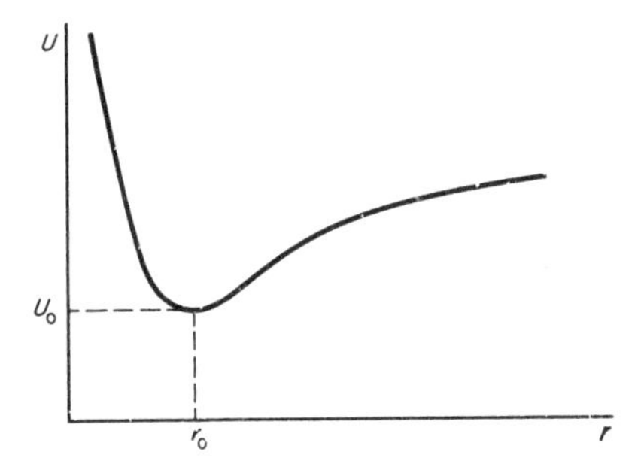

FIG. 5.2. Relationship between U, the internal energy of a system consisting of two ions of opposite charges, and r, their interionic distance. R_0 is the equilibrium distance between the ions, U_0 the "lattice energy" of the crystal of which they are part – the internal energy of the crystal in its stable state.

atoms in its otherwise neutral covalent molecule. This causes an association between the water molecules, known as a "hydrogen" bond (Fig. 5.3), and the result is a liquid of high

FIG. 5.3. Two polar water molecules showing the "hydrogen" bond between the slightly negatively charged oxygen atom of one molecule and a slightly positively charged hydrogen atom of the other.

dielectric constant which is able to cause the breakdown of the crystal lattice of ionic compounds because the attractive force between the two unlike charges is inversely proportional to the dielectric constant of the medium between them.

$$F = \frac{Q_1 Q_2}{\epsilon r^2}, \tag{5.1}$$

where F is the attractive force between unlike charges Q_1 and Q_2, r is the distance between the charges, and ϵ is the dielectric constant of the medium between the charges.

When the salt MX is dissolved in water, each ion will have a *heat of solvation*, ΔH_{M^+} and ΔH_{X^-} for the positive and negative ions respectively. These heats of solvation represent the difference in enthalpy between the gaseous ion and the ion in solution. The process of solution of the salt MX in water can then be represented by a Born–Haber cycle[1]

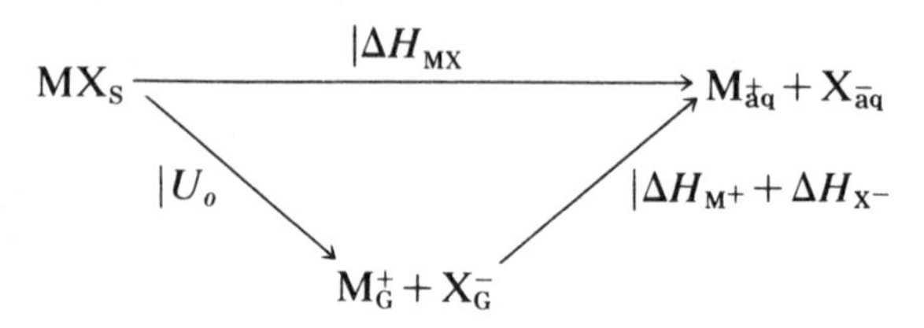

This cycle is an application of the First Law of Thermodynamics (see Chapter 1), as the heat of solution of the salt ΔH_{MX} can be determined by representing the process as taking place in two stages,

I. Break-up of the lattice of the solid MX to give isolated gaseous ions, accompanied by an absorption of the lattice energy U_o.

$$MX_S \xrightarrow{|+U_o} M^+_G + X^-_G$$

II. Solvation of the gaseous ions in water, accompanied by the evolution of their heats of solvation.

$$M^+_G \xrightarrow{|+\Delta H_{M^+}} M^+_{aq}$$

$$X^-_G \xrightarrow{|+\Delta H_{X^-}} X^-_{aq}$$

Then the heat of solution ΔH_{MX} is the sum of the energy changes accompanying stages I and II,

$$\Delta H_{MX} = U_o + \Delta H_{M^+} + \Delta H_{X^-}. \qquad (5.2)$$

ΔH_{MX} may be positive or negative depending on the relative magnitudes of the terms on the right-hand side of the equation, but is usually positive in the case of ionic solids dissolving in water. This is the basis of "freezing mixtures" which depend on the endothermic nature of the process of solution of ionic compounds.

In addition to its ability to break up ionic lattices, water is a *ligand*, because it forms complexes with metal ions or atoms. A ligand is an ion or molecule which has at least one pair of unshared electrons (a "lone pair") which is donated to the metal atom or ion to form a *co-ordinate bond* by satisfying the metal's electronic requirements, most commonly the empty *d* orbitals of the transition metals.[2] Figure 5.4 shows a cupric

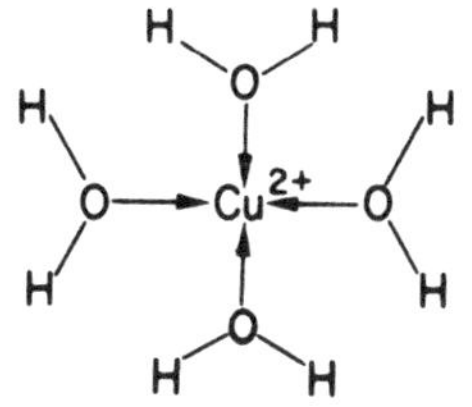

FIG. 5.4. Hydrated cupric ion, $[Cu(H_2O)_4]^{2+}$. Co-ordinate bonds represented by arrows → where the oxygen atoms from each of the four water molecules donate their "lone pairs" of electrons to the cupric ion.

ion forming a complex with four water molecules. Each water molecule supplies a "lone pair" of electrons which are taken up in the orbitals of the cupric ion, Cu^{2+}. The water molecules form a "solvation sheath" round the metal ion.

Water is one of many ligands, which contain atoms with "lone pairs" of electrons suitable for forming co-ordination complexes with metal ions. The atoms in ligands which are bound directly to the metal ions in the complexes are those of the more electronegative elements (e.g. C, N, P, O, S, F, Cl) and examples of ligands include the following

$$NH_3, Cl^-, CN^-, OH^-, P_2O_7^{4-}.$$

The complex formed by these ligands may be more or less stable than that formed by the water, and we would therefore see variations in the heat of solvation of the metal ion according to the complexing ligands present in the solvation sheath, as the energy involved in the formation of the solvated ion will be included in the term ΔH_{M^+} of (5.2). An example of the introduction of other ligands into an aquo-ion is seen in the addition of HCl to an aqueous solution of cupric sulphate. The blue colour due to the aquo-ion $[Cu(H_2O)_4]^{2+}$ changes to green as the chloride ions enter the solvation sheath, progressively replacing the water molecules, and producing a negatively charged complex ion.[2]

$$[Cu(H_2O)_4]^{2+} \rightarrow [Cu(H_2O)_3Cl]^{+} \rightarrow [Cu(H_2O)_2Cl_2] \rightarrow$$
$$\rightarrow [Cu(H_2O)Cl_3]^{-} \rightarrow [CuCl_4]^{2-}.$$

In the reaction

$$[Cu(H_2O)_4]^{2+} + Cl^{-} = [Cu(H_2O)_3Cl]^{+} + H_2O,$$

the equilibrium constant of the reaction can be used to calculate the *stability constant* of the complex ion $[Cu(H_2O)_3Cl]^{+}$.

We now have a picture of aqueous solutions of ionic compounds containing solvated metal ions instead of uncomplexed metal ions and water molecules, and if there is to be any motion of these comparatively large constituents, the hindrance to their motion will be much greater than might be expected from the simple model which ignores the existence of such complexes.

The problems of high-temperature experimental work have delayed the accumulation of detailed knowledge of the structure and properties of *molten salts*, such as is available for aqueous solutions. Nevertheless, it is known that salts which are ionic in the solid state have a similar structure when they are molten and within a few hundred degrees of their melting points. They have a short-range structure similar to the solid, but the long-range crystal structure is lost. They have similar densities and specific heats, and their entropies of fusion are small—demonstrating that there has been no major change in the state of order of system on fusion. They appear to have considerably more vacant lattice sites than their equivalent solid structure, and this should enable diffusion to take place more easily than in solids (see Section 4.12). Simple molten salts are good electrolytic conductors—better in fact than aqueous solutions of salts—and the conduction is wholly ionic, showing that they have an ionic structure rather than a molecular structure. Bloom and Bockris, in a comprehensive review,[(3)] list the following among the properties of molten electrolytes which have been investigated—heats and entropies of fusion, molar volume, electrical conductance, viscosity, vapour pressure, compressibility, refractive index, diffusion, transport number (see Section 5.4), cell e.m.f.'s, surface tension and heat capacities. Spectroscopic and X-ray diffraction investigations have also been carried out.

The picture of the structure of *molten simple salts* (KCl, $PbCl_2$, $CdCl_2$, NaF, CdI_2, KI, $MgCl_2$, $BaCl_2$, for example) is generally one of ions such as Na^+, Cl^-, K^+, etc., possibly some partially undissociated molecules ($ZnCl_2$ in $NaNO_3$), vacancies, and complex ions ($[CdI_4]^{2-}$ in CdI_2–KI mixtures and $[AlF_6]^{3-}$ in the AlF_3–NaF system; this latter system contains cryolite, the solvent for Al_2O_3 used in the electrolytic extraction of aluminium). Both positive and negative ions appear to move in the conduction of electricity in molten salts and electrolysis of molten halides is an important source of metals such as magnesium, calcium, sodium, beryllium and niobium.

Metal oxides form conducting melts of much greater complexity than the simple halides. They include silicates, phosphates, borates and aluminates, which form the basis of metallurgical slags, glasses and ceramics. Unlike the halides they have a tendency to form long-range giant ions by covalent bonding, polymerizing in an analogous way to the compounds of carbon. These large ions produce a highly viscous melt, and are often based on silica, which is a compound distributed abundantly in naturally occurring ores and rocks. Most work has been carried out on mixtures of oxides rather than the pure oxides, with an important emphasis on the RO–SiO_2 systems where R may be Ca, Fe, Mg, Mn, Pb, Zn, etc.

Largely ionic conduction of a lower magnitude than in simple salts is indicated, but often the main contributor is the relatively small positive ion R^{2+} which is much more mobile than the large polymer negative ions $[Si_xO_y]^{n-}$. Pure silica, even in the liquid state, consists of a three-dimensional network with relatively few broken Si–O bonds. Each silicon atom is attached to four oxygen atoms, each oxygen atom to two silicon atoms (Fig. 5.5). When a metal oxide RO is added to the silica, silicon–oxygen bonds are broken (Fig. 5.6) as the oxygen ion attaches itself to the silica network to form a negative silicate ion. The metal ion is completely surrounded by the silicate network, and the difference between this and the structure of molten halides is shown by the high activation energy (290 kJ/mol) for diffusion of calcium in lime–alumina–silica melts compared with a value of 17 kJ/mol for sodium in molten sodium chloride.[4] Electrical neutrality is

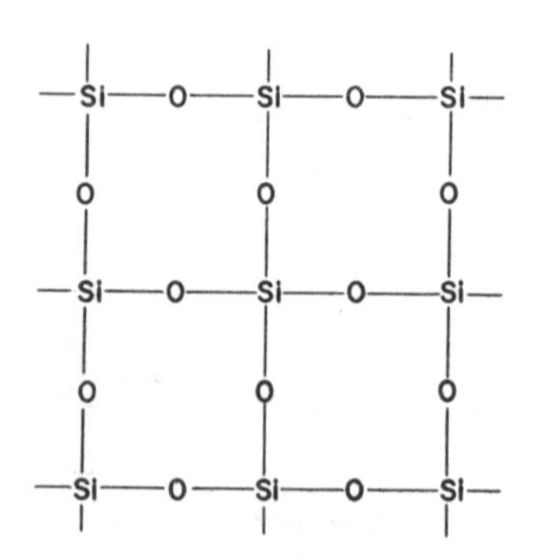

FIG. 5.5. Representation in two dimensions of part of the three-dimensional network of silica.

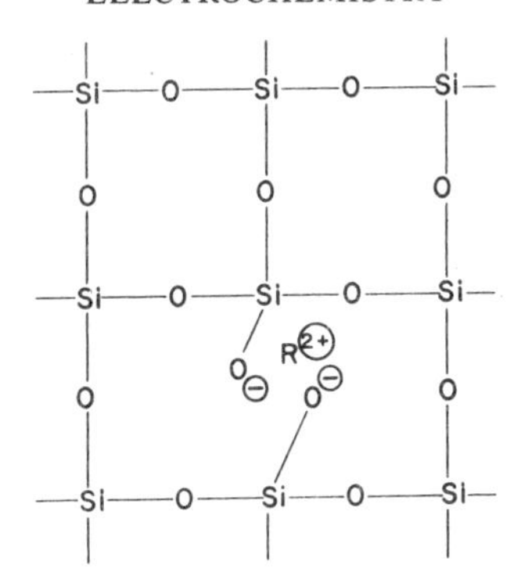

FIG. 5.6. Si – O – Si bond in silica network broken to accomodate the metal oxide RO. A negative silicate ion of large size is produced, electrically balanced by the presence of the positive ion, R^{2+}.

maintained between the positive metal ions and the large negative silicate ions as more of the oxide RO is added and more Si – O – Si bonds are broken. Eventually, at the composition in which there are two RO molecules for every SiO_2 molecule, all the Si – O – Si bonds will be broken and the silicon will be entirely present as $[SiO_4]^{4-}$ ions, and the metal R as R^{2+} ions. This is the "orthosilicate" composition, a composition of minimum viscosity because of the complete absence of polymer silicate ions. Further additions of RO produce free oxygen ions O^{2-} in the structure.

The addition of oxides to halide melts produces a very complex liquid structure, an example being Al_2O_3 added to AlF_3-NaF mixtures in the electrolyte used for electrolytic extraction of aluminium. The Al^{3+} ion seems to be part of some sort of complex with O^{2-} and F^- distributed around it depending on the composition of the solution, with no clear ionic species present, though $[AlO_2F_2]^{3-}$, $[AlO_2]^-$ and $[AlF_6]^{3-}$ have been suggested.

The perfect crystal structure of the *solid* ionic salt MX represented in Fig. 5.1 is rarely the structure found in nature.[5] Instead, vacancies and interstitials (Schottky and Frenkel defects respectively – see Section 4.13) may be present (Fig.

5.7). The salt of the structure shown in Fig. 5.7a would still have the stoichiometric composition MX as there is one X^- ion defect for each M^+ ion defect, and an example of the structure in Fig. 5.7b is AgBr which contains Frenkel defects

M^+ X^- M^+ X^- M^+ X^- M^+
X^- □ X^- M^+ X^- M^+ X^-
M^+ X^- M^+ □ M^+ X^- M^+
X^- M^+ X^- M^+ X^- M^+ X^-

(a)

M^+ X^- M^+ X^- M^+ X^- M^+ X^-
M^+
X^- M^+ X^- M^+ X^- M^+ X^- M^+
X^-
M^+ X^- □ X^- M^+ □ M^+ X^-
X^- M^+ X^- M^+ X^- M^+ X^- M^+

(b)

FIG. 5.7. (a) Schottky defects (vacancies □) in M^+ and X^- lattices of the solid salt MX. (b) Frenkel defects (interstitials) near vacancies in the M^+ and X^- lattices of the solid salt MX.

in the cation lattice (Fig. 5.8). Because the anion is usually larger than the cation, Frenkel defects in anion lattices are unlikely. Where some of the M^+ ions, or alternatively the X^- ions, are missing, we will have a non-stoichiometric compound $MX_{1+\delta}$ or $M_{1+\delta}X$. An interesting example of a defect lattice is the mixed oxide of zirconia (ZrO_2) and 14 mole

Ag^+ Br^- Ag^+ Br^- Ag^+ Br^-
Br^- Ag^+ Br^- Ag^+ Br^- Ag^+
Ag^+
Ag^+ Br^- □ Br^- Ag^+ Br^-
Br^- Ag^+ Br^- Ag^+ Br^- Ag^+

FIG. 5.8. Representation of the AgBr lattice, showing a Frenkel defect in the Ag^+ lattice.

% magnesia (MgO) which contains vacancies in the oxygen ion lattice—the oxide is called "doped zirconia" (Fig. 5.9a). Wüstite, nominally FeO, has a non-stoichiometric structure approximating to $Fe_{0 \cdot 95}O$ at 600°C, and has excess oxygen ions (Fig. 5.9b). We saw in Section 4.13 that the presence of vacancies and interstitials made diffusion in the solid state possible. Defect lattices in solid salts allow the diffusion of ions under the influence of an electrical potential and hence

Zr^{4+} O^{2-} Zr^{4+} O^{2-} Zr^{4+}
O^{2-} Mg^{2+} O^{2-} Zr^{4+} O^{2-}
Zr^{4+} □ Zr^{4+} O^{2-} Zr^{4+}
O^{2-} Zr^{4+} O^{2-} Zr^{4+} O^{2-}

(a)

Fe^{2+} O^{2-} Fe^{2+} O^{2-} Fe^{2+}
O^{2-} Fe^{2+} O^{2-} Fe^{3+} O^{2-}
Fe^{2+} O^{2-} □ O^{2-} Fe^{2+}
O^{2-} Fe^{3+} O^{2-} Fe^{2+} O^{2-}
Fe^{2+} O^{2-} Fe^{2+} O^{2-} Fe^{2+}

(b)

FIG. 5.9. (a) Representation of the lattice of zirconia "doped" with magnesia; there is a vacancy in the oxygen lattice. (b) Vacancy in the Fe^{2+} lattice of wüstite, which contains two ferric ions to balance the loss of one ferrous ion.

electrolysis in the solid state. They are also important in the structure and behaviour of catalysts, of oxide films on metals, semiconductors, and in gas–solid reactions such as the reduction of iron oxide by carbon monoxide in the iron blast furnace, and have become the basis of interesting fundamental work with solid state electrolysis.[6]

Unfortunately, electronic conductivity becomes significant at high temperatures, otherwise doped zirconia might have been a suitable electrolyte for the measurement of oxygen activities in steel and steelmaking slags.

The structure of metals is one of metal ions forming a lattice in which the metals' valency electrons can be pictured as a "cloud" permeating the whole lattice. This simplified picture can be used to account for the electronic conduction of metals. However, if ions are present in a metallic structure, some electrolytic conduction might be expected even though it forms a very small proportion of the total conduction of a metal. Electrotransport of oxygen in solid β-titanium and zirconium has been studied, with the oxygen migrating towards the positively charged end of the electrolyte. Nitrogen behaves similarly in γ-iron. Carbon migrates to the negative electrode in α- and γ-iron. In the β-phase of the Cu–Al system, the aluminium migrates to the positively charged electrode.

Electrolytic transport in liquid metals and alloys is more rapid, as might be expected from the more loosely bound structure and greater vacancy content of liquids. Reference 7 should be consulted for further discussion of electrotransport in metals.

5.3. Electrolysis

If two *electronic* conductors, called "electrodes", are placed in contact with an electrolyte consisting of ions X^+ and Y^-, and an electrical potential is applied to the electrodes—for example, by means of a battery—then the positively charged ions X^+ will be attracted by the negatively charged electrode, and the negative ions Y^- by the positively charged electrode. The positively charged ions X^+ are called *cations* and the negatively charged ions Y^- *anions*. The electrode to which the *cations* are attracted is called the *cathode*, and the elec-

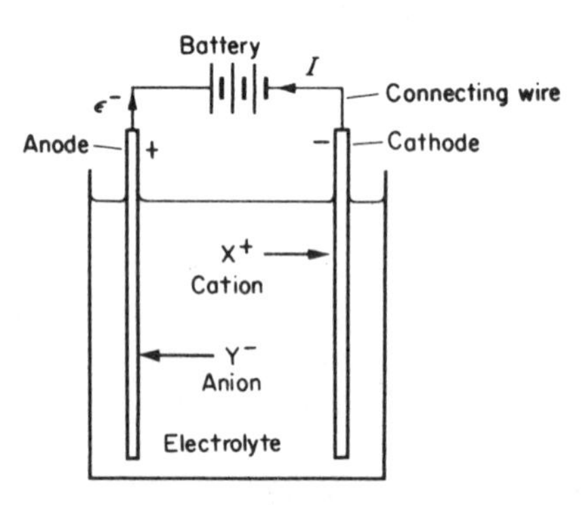

Fig. 5.10. Electrolysis. I shows the direction of conventional current flow, ϵ^- the direction of the electron flow in the electronic conductor.

trode to which the *anions* are attracted is called the *anode* (see Fig. 5.10). This process, known as *electrolysis*, occurs whether the electrolyte is an aqueous solution or a molten salt, whether the ions are simple or complex.

The cations, when they reach the cathode, show a tendency to pick up electrons, forming an electrically neutral species X, which "discharges" at the cathode,

$$X^+ + \epsilon^- = X.$$

The anions lose electrons and discharge as the electrically neutral species Y on reaching the anode,

$$Y^- = Y + \epsilon^-.$$

We can now define a *cathode* as an electrode at which electrons tend to be consumed by the discharging species, and an *anode* as an electrode at which electrons tend to be produced as a result of the electrode reaction. In this process electrons flow from the anode to the cathode via the battery—in the opposite direction to the conventional "current" I.

An example of this behaviour is the electrolytic extraction of magnesium from an electrolyte of molten magnesium chloride in solution with molten sodium chloride, potassium chloride and calcium chloride at 750°C (the I.G. Farben process). The cations Mg^{2+} are attracted to the steel cathode, and the anions Cl^- to the graphite anode. Both ions discharge, producing molten magnesium and chlorine gas respectively.

Cathode reaction. $Mg^{2+} + 2\epsilon^- = Mg_L.$
Anode reaction. $2Cl^- = Cl_{2G} + 2\epsilon^-.$

This is a very simple representation of electrode reactions, which can often be much more complex, as will be seen later in this chapter. However, it expresses the net result of the electrolytic process, which is the discharge of the species Mg and Cl_2 at the electrodes, and the transfer of electrons from anode to cathode.

M. Faraday studied the electrolytic decomposition of aqueous electrolytes, and published (1834) the following laws relating the quantity of electricity passed (the product of the current and the time during which it was passed) to the amount of substance discharged at an electrode:

Faraday's First Law of Electrolysis: the amount of any substance discharged or dissolved at an electrode is proportional to the quantity of electricity passed.

Faraday's Second Law of Electrolysis: if the same quantity of electricity is passed through a variety of electrolytes, the amounts of different substances discharged or dissolved at the electrodes is proportional to the chemical equivalent weights of those substances.

Examination of the electrode reactions in the magnesium chloride electrolysis mentioned above will show why these laws apply. The flow of an electric current is the flow of electrons – and as each electron is produced or taken up at an electrode, it will cause the discharge of a definite amount of magnesium or chlorine. In the case of magnesium, two electrons cause the discharge of one magnesium ion. The equivalent weight of magnesium is half its atomic weight. Two electrons given up by chlorine ions at the anode cause the discharge of two chlorine ions. The equivalent weight of chlorine is equal to its atomic weight, and Faraday's laws are obeyed in both cases.

The unit quantity of electricity is the *coulomb*, which is the quantity of electricity passed when a current of *one ampere* flows for *one second*. The quantity of electricity required to deposit one gramme equivalent of a substance (the equivalent weight of the substance expressed in grammes) is called the *faraday* F and is 96,500 coulombs within the limits of experimental accuracy normally required. If an ion of valency z is being discharged electrolytically, each gramme ion will contain z gramme equivalents, and will require zF coulombs to cause its discharge. In each gramme ion there are N ions, where N is Avogadro's number ($6{\cdot}023 \times 10^{23}$), so that the charge of an electron – the unit electrical charge – is

$$e = -\frac{F}{N} = -\frac{96{,}500}{6{\cdot}023 \times 10^{23}} = -1{\cdot}60 \times 10^{-19} \text{coulombs}.$$

If for any reason—such as the consumption of electrical power in heating the electrolyte by "ohmic heating"—the amount of substance discharged at an electrode is less than that predicted by Faraday's laws, the *current efficiency* at the electrode is defined as

$$\frac{\text{Actual amount discharged}}{\text{Theoretical amount discharged}} \times 100.$$

This factor is important in the economic assessment of an electrolytic process. Examples of typical current efficiencies are:

	Cathode current efficiency
Electrolytic extraction of copper from aqueous solution	80%
Electrolytic refining of copper in aqueous solution	95%
Electrolytic extraction of aluminium from fused salts	90%
Electrolytic extraction of zinc from aqueous solution	90%

5.4. Conduction in Electrolytes

Ohm's law is obeyed by electrolytic conductors, where

$$V = IR, \tag{5.3}$$

V being the potential difference applied across the conductor, I the current flowing, and R the resistance of the conductor. Increasing temperature increases the rate of diffusion of the ions in the electrolyte, and therefore decreases the resistance—in contrast to the increase in resistance of electronic conductors with temperature. The resistance of electrolytes can be measured by means of a Wheatstone bridge circuit, using an alternating current source of e.m.f. to prevent "polarization" of the electrodes (see Section 5.12). Reference should be made to textbooks of physical chemistry[8] for experimental details. As a result of such measurements, the following properties of an electrolyte can be determined:

Specific Resistance (or *Resistivity*) ρ: the resistance in ohms of a portion of the electrolyte 1 cm long and 1 cm² in cross-sectional area. (Units: ohm cm.)

$$\rho = \frac{RA}{l}, \tag{5.4}$$

where A = cross-sectional area in square centimetres, l = length of electrolyte in centimetres.

Specific Conductance (or *Conductivity*) κ: the reciprocal of the specific resistance. (Units: ohm^{-1}/cm, mho/cm, reciprocal ohm/cm and the SI unit is the siemen per cm.)

$$\kappa = \frac{l}{RA} \tag{5.5}$$

Equivalent Conductance Λ: the conductance of that volume of the electrolyte which contains one gramme equivalent of the ions taking part in the electrolysis and which is held between parallel electrodes one centimetre apart. (Units: ohm^{-1} cm^2.)

If v cubic centimetres is the volume of the solution containing one gramme equivalent, then the value of l will be 1 cm and the value of A will be v square centimetres, so that

$$\Lambda = \kappa \frac{A}{l} = \kappa v. \tag{5.6}$$

In aqueous solutions, concentrations are sometimes expressed in terms of "normality" (gramme equivalents per litre), so that if c is concentration, then

$$v = \frac{1000}{c},$$

and [from (5.6)]

$$\Lambda = \frac{1000\,\kappa}{c}. \tag{5.7}$$

The equivalent conductance is used because it is the measure of the electrolytic conductance of the ions which make up 1 g-equiv. of electrolyte of a particular concentration —thereby setting conductance measurements on a common basis. Sometimes the *molar conductance* Λ_m is preferred to the equivalent conductance; it is the conductance of that volume of the electrolyte which contains one gramme molecule (mole) of the ions taking part in the electrolysis and which is held between parallel electrodes 1 cm apart.

The equivalent conductance of an aqueous electrolyte varies with concentration according to the modified form of the Onsager equation (1927)[9]

$$\Lambda_c = \alpha[\Lambda_o - (A + B\Lambda_o)\sqrt{(\alpha c)}]. \tag{5.8}$$

Here A and B are constants depending on temperature, viscosity of the solvent, and dielectric constant of the solvent, c is the concentration in gramme equivalents per litre, and Λ_c is the equivalent conductance of the solution. Λ_o is the *equivalent conductance at infinite dilution*—that is, at $c = 0$, when the ions are infinitely distant from one another and there is no interionic attraction. α is the *degree of dissociation* of the electrolyte. For example, with the compound XY,

$$XY = X^+ + Y^-,$$

and a fraction α of the molecules XY may be dissociated into ions X^+ and Y^-. With "strong electrolytes"—such as salts and strong acids in water—the substances are known to be "completely ionised" (Section 5.2), so that $\alpha = 1$ and from (5.8),

$$\Lambda_c = \Lambda_o - (A + B\Lambda_o)\sqrt{c}, \tag{5.9}$$

which is the original form of the Onsager equation.

With weak electrolytes, such as acetic acid and ammonia in water, α is less than 1, and (5.8) is obeyed. It must be stressed that (5.8) and (5.9) are only obeyed in fairly dilute solutions. The relationships are illustrated in Fig. 5.11, showing the large increase in Λ_c for weak electrolytes when c approaches zero.

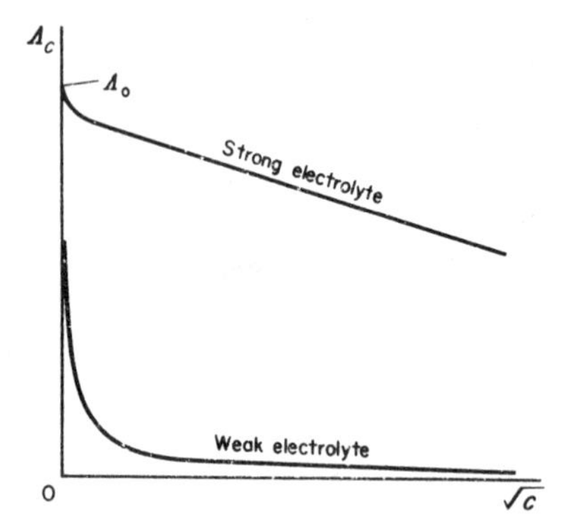

FIG. 5.11. Relationship between the equivalent conductance Λ_c and the concentration c for strong and weak electrolytes. Λ_0 is the equivalent conductance at infinite dilution for the strong electrolyte.

F. W. Kohlrausch (1875) found that, at infinite dilution, each ion in the electrolyte contributes a characteristic amount to the equivalent conductance of the electrolyte, so that for the electrolyte containing the salt XY

$$\Lambda_0 = \Lambda_{X_+} + \Lambda_{Y_-}, \tag{5.10}$$

where Λ_{X_+} and Λ_{Y_-} are the *equivalent ionic conductances* (at infinite dilution). Each ion has its own ionic conductance which it displays whenever it appears in the same solvent. It must be remembered that by "ion", we mean "ionic complex"–so that if, for example, the ligands forming a complex ion with a metal ion in aqueous solution change, the ionic conductance will change because its mobility in the solution will have altered. Hydrogen and hydroxyl ions in aqueous solution have high ionic conductances, whereas the ionic conductance of metal ions decreases as the tendency to form complexes–and therefore more bulky ions–increases.

Glasstone[1] gives the following values (expressed in ohm^{-1}/cm^2) for aqueous solutions at 25°C, $\Lambda_{H+} = 349{\cdot}82$, $\Lambda_{OH-} = 198{\cdot}5$, $\Lambda_{K+} = 73{\cdot}52$, $\Lambda_{Na+} = 50{\cdot}11$, $\Lambda_{Li+} = 38{\cdot}69$, $\Lambda_{Cl-} = 76{\cdot}34$.

Although Li^+ is a small ion, it shows a greater tendency to solvation, so that its ionic conductance in water is smaller than that of larger ions like Na^+.

Measurements of equivalent conductance in molten salts show that some salts (e.g. $ZnCl_2$) are incompletely dissociated ($\alpha < 1$). Mixtures of molten salts should, ideally, obey the following equation

$$\Lambda_{\text{mixture}} = x_A \,.\, \Lambda_A + x_B \,.\, \Lambda_B, \tag{5.11}$$

where x_A and x_B are the mole fractions of the salts A and B respectively, and Λ_A and Λ_B are the equivalent conductances of pure A and pure B. However, deviations from this ideal equation are experienced,[3] and this may indicate the presence of complex ions (e.g. CdI_4^{2-} in CdI_2–KI mixtures) where the deviation is negative,

$$\Lambda_{\text{mixture}} < x_A \,.\, \Lambda_A + x_B \,.\, \Lambda_B.$$

A positive deviation (e.g. in $MgCl_2 - CaCl_2$) would indicate that the association between ions is less in the mixture than in the pure state. Reasoning along these lines could lead to concepts of "activities" in molten electrolytes in an argument similar to those in Sections 3.3 and 3.4, where the relative attraction between components of a solution was used to explain deviations in vapour pressure measurements from Raoult's law. The subject of activities will be taken up again later in this chapter.

The velocity of an ion transported electrolytically down a potential gradient (the potential difference divided by the distance between the electrodes) of 1 V/cm is called the *ionic mobility* or *velocity*, u_+ for cations, u_- for anions.

$$\begin{aligned} \text{Then } \Lambda_0 &= \Lambda_+ + \Lambda_- \qquad \text{(from 5.10)} \\ &= k(u_+ + u_-), \end{aligned} \tag{5.12}$$

where k is a constant. Thus

$$\Lambda_+ = ku_+, \quad \Lambda_- = ku_-. \tag{5.13}$$

Ohm's law (5.3) shows that the current is inversely proportional to the resistance—and therefore directly proportional to the conductance. Thus

$$I \propto u_+ + u_- \quad \text{[from (5.12)]}$$

and that fraction of the current carried by a particular ion is proportional to its ionic mobility u. This fraction of the total current carried by a particular ion is called the *transport number* (t) of the ion, where

$$t_+ = \frac{u_+}{u_+ + u_-}, \quad t_- = \frac{u_-}{u_+ + u_-}. \tag{5.14}$$

Measurements of transport numbers of ions in electrolytes can give important information on the mechanism of conduction and on the nature of the ions involved. In silicate melts, at compositions containing more silica than the orthosilicate ($2RO, SiO_2$), the transport number of the cations is almost unity—indicating that most of the current is carried by the relatively free cations, and very little by the large polymer silicate anions.[3] References 1 and 3 should be consulted for a discussion of the experimental determination of transport numbers in aqueous solutions and molten salts respectively.

5.5. The Thermodynamics of the Reaction at an Electrode

We will now consider the chemical reaction at an electrode, and in this treatment we can define an electrode as an electronic conductor in contact with an electrolytic conductor. In the case of the reaction at a metal electrode in which ions of the same metal "discharge" at the electrode from the electrolyte, the reaction can be represented by

$$M^{z+} + Z\epsilon^- = M. \tag{5.15}$$

Depending on the electrolyte, whether it is aqueous, solid or a fused salt (see Section 5.2), the metal ions will form part of the structure of that electrolyte. They may be present as aquo-ions—surrounded by a solvation sheath of water molecules (Fig. 5.4)—complex ions such as the complex formed between cyanide ions and cupric ions, or completely surrounded by a silicate network (Fig. 5.6) in liquid slags. The metal electrode will consist of metal ions in a crystal structure bound together by the attraction of their "free electrons" which give the metal its electronic conductivity. We can say that the metal ions in the electrolyte possess a free energy G_e and the metal ions in the electrode possess a free energy G_m. Then, if the metal ion is to leave its place in the electrolyte structure and take up a position in the structure of the metal electrode, the free energy change accompanying this process will be

$$\Delta G = G_m - G_e, \tag{5.16}$$

and the reaction (5.15) could be written

$$M_e^{z+} = M_m^{z+}, \tag{5.17}$$

where M_e^{z+} is the metal ion in the electrolyte, M_m^{z+} the metal ion in the metallic electrode. We know from the van't Hoff isotherm (2.37) and (2.39) that the free energy change ΔG accompanying this reaction will be given by

$$\Delta G = \Delta G^{\circ} + RT \ln \frac{c_{M_m^{z+}}}{c_{M_e^{z+}}} \tag{5.18}$$

where 1 mole of ions is involved.

If the concept of activities developed in Chapter 3 is used, then, to account for deviations from ideal solutions should they occur, we should replace concentrations by activities in (5.18)

$$\Delta G = \Delta G^{\circ} + RT \ln \frac{a_{M_m^{z+}}}{a_{M_e^{z+}}}. \tag{5.19}$$

The activity of a metal ion in a pure metal $a_{M_m^{z+}}$ can be considered to be unity if the pure metal is its standard state, and we can write $a_{M^{z+}}$ for the activity of the metal ions in the electrolyte. Changing the sign of the logarithmic term, (5.19) can now be written

$$\Delta G = \Delta G^{\circ} - RT \ln a_{M^{z+}}, \tag{5.20}$$

the free energy change accompanying the discharge of one gramme ion of the metal ion M^{z+} at an electrode of the metal M by (5.15).

As this is a chemical reaction, we would expect the reactants to have to go through some activated transition state (Section 4.8), involving a free energy of activation ΔG^* (Section 4.9). The free energy changes taking place in this process can be represented in diagrams such as those in Fig. 5.12, where the energy–distance curve is plotted for a negative value of the free energy change accompanying the electrode reaction [$G_e > G_m$ – see (5.16)]. If G_e were less than G_m, ΔG would

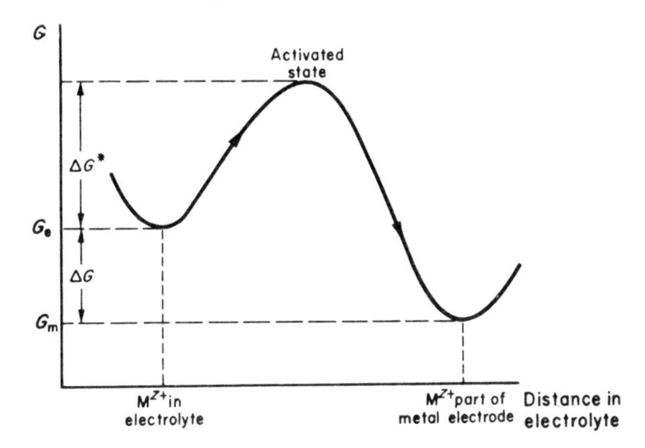

FIG. 5.12. Free energy G of metal ion M^{z+} as it travels from the electrolyte to discharge at an electrode of the metal M. ΔG is negative for this process which is consequently spontaneous. ΔG^* is the free energy of activation, G_m and G_e the free energies of the ion in the metal and electrolyte structures respectively.

be positive and the metal M would tend to enter the electrolyte as the ion M^{z+}.

5.6. The Electrode Potential: The Galvanic Cell

Clearly, reactions such as (5.15) will have a driving force, depending on the value of ΔG, and a measure of that driving force will give a measure of the tendency of the metal ion to discharge at the metal electrode or for the metal ion of the electrode to enter the electrolyte. An alternative approach is to consider that there is an electrical potential difference E between the ion's position in the electrolyte and its position in the structure of the metal electrode. The work done in moving unit electrical charge through such a potential difference is equal to E units of electrical potential, by definition. The charge associated with one gramme ion M^{z+} is zF coulombs, where F is the Faraday, 96,500 coulombs, (Section 5.3), so that, if the potential difference between electrolyte and electrode is E volts, the work done w in the transfer of a charge of zF coulombs through this potential difference is EzF joules,

$$w = EzF. \tag{5.21}$$

Gibbs (1875) and von Helmholtz (1882) realized that this electrical work done was equal, at constant pressure, to $-\Delta G$, the free energy change accompanying the process, by a similar argument to that leading to (2.17)

$$\Delta G = -w + P\Delta V.$$

There is no measurable volume change accompanying this process, so that $\Delta V = 0$ and, using (5.21),

$$\Delta G = -EzF. \tag{5.22}$$

This is an important relationship, and the units of the terms involved should be remembered. ΔG is in joules, z is the valency exhibited by the ion M^{z+} in the electrolyte, and F is the Faraday, 96,500 coulombs. E is measured in volts and is known as the *electrode potential* of the electrode.

Substituting for ΔG in (5.20),

$$-EzF = -E^{\circ}zF - RT \ln a_{M^{z+}},$$

or, dividing through by $(-zF)$,

$$E = E^{\circ} + \frac{RT}{zF} \ln a_{M^{z+}}, \tag{5.23}$$

where E° is called the *standard electrode potential* of the electrode. Equation (5.23) is known as the *Nernst Equation*, after W. Nernst (1889). When the ions M^{z+} are in their standard state of unit activity, $\ln a_{M^{z+}} = 0$ and E is equal to the standard electrode potential E°. It must be remembered that equations such as (5.18), (5.20), and (5.23) refer only to *thermodynamically reversible* processes (Section 1.5). This means that E is the *reversible electrode potential*, and only where the electrode is behaving thermodynamically reversibly can it apply.

If two electrodes of different metals M_1 and M_2 are placed in a common electrolyte containing ions of both metals, $M_1^{z_1+}$ and $M_2^{z_2+}$ respectively, both electrodes will have an electrode potential, E_1 and E_2 respectively (Fig. 5.13). The Nernst equation (5.23) gives the values of E_1 and E_2:

$$E_1 = E_1^0 + \frac{RT}{z_1F}\ln a_{M_1^{z_1+}},$$

$$E_2 = E_2^0 + \frac{RT}{z_2F}\ln a_{M_2^{z_2+}}.$$

These electrode potentials measure the tendency of the metal ions to discharge at the electrodes, and in the last section we saw that the more positive the electrode potential, the more negative ΔG, the free energy change accompanying the discharge process. This means that the higher the value of E, the greater the tendency for the metal ion to leave the electrolyte and discharge at the metal electrode. Thus if $E_2 > E_1$, metal ions $M_2^{z_2+}$ will have a greater tendency to dis-

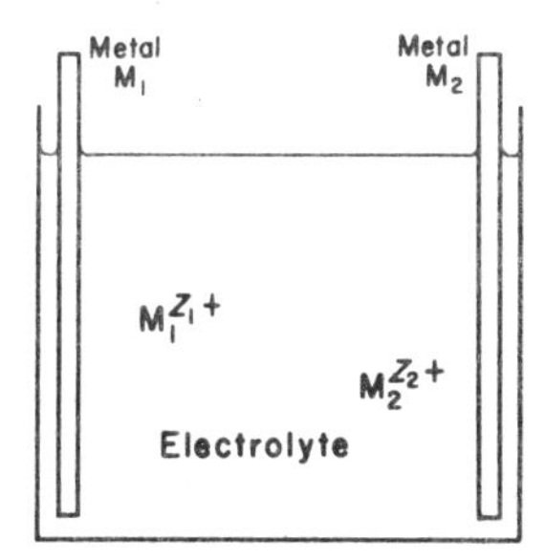

FIG. 5.13. Dissimilar electrodes of metals M_1 and M_2 in an electrolyte containing ions $M_1^{z_1+}$ and $M_2^{z_2+}$.

charge at the metal electrode M_2 than $M_1^{z_1+}$ ions at the M_1 electrode.

If the two electrodes were left in the condition shown in Fig. 5.13, no electrode reactions could take place because the electrons in the metal electrodes would have no complete electronic conducting circuit along which they could flow. If the two electrodes were connected outside the electrolyte by an electronically conducting wire (Fig. 5.14), such a circuit

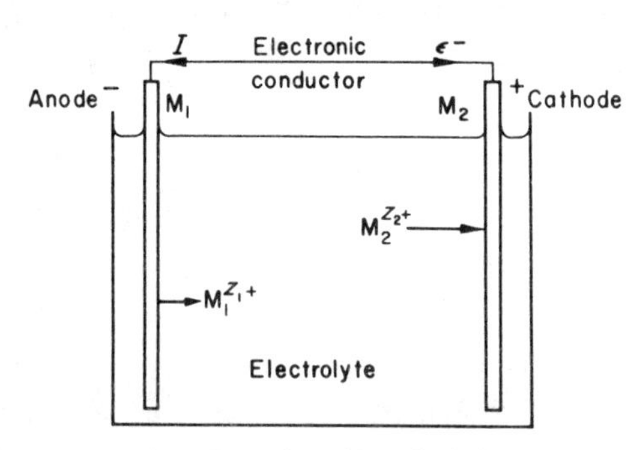

FIG. 5.14. Galvanic cell produced by dissimilar metal electrodes M_1 and M_2 joined by an electronic conductor and dipping into an electrolyte. M_1 goes into solution at the anode, M^{z_2+} ions discharge at the cathode.

would be provided. Because E_2 was greater than E_1, the electrode process at electrode M_2 would have a greater tendency to progress and this reaction would take place, as in (5.15),

$$M_2^{z_2+} + z_2\epsilon^- = M_2. \tag{5.24}$$

Electrons would be used up in this process and would have to be produced by a net electron flow along the connecting wire from the electrode of metal M_1. In order that electrons be produced at the electrode M_1, the electrode reaction at M_1 would have to be reversed

$$M_1 = M_1^{z_1+} + z_1\epsilon^-. \tag{5.25}$$

The greater driving force ΔG_2 of the reaction at M_2 would therefore be driving the reaction at M_1 contrary to its natural tendency, and the free energy change at M_1 would be $-\Delta G_1$. The sum of reactions (5.24) and (5.25) would give the overall reaction of the "galvanic" cell produced,

$$M_2^{z_2+} + M_1 = M_1^{z_1+} + M_2. \tag{5.26}$$

The free energy change ΔG accompanying this overall cell reaction would be given by the sum of the free energy changes occurring at the two electrodes, and would produce an "electromotive force" (e.m.f.), E_{cell}.

$$\Delta G = \Delta G_2 - \Delta G_1, \tag{5.27}$$

and, from (5.22)

$$\Delta G = -E_{cell}\, zF,$$

where z must be the same as z_1 and z_2 to preserve an electron balance. (If the metal ions are of different valency, there must be an adjustment in the amount of metal M_2 discharged and M_1 taken into the electrolyte according to Faraday's laws of electrolysis, for example

$$3Cu^{2+} + 2Cr = 3Cu + 2Cr^{3+}.)$$

We will therefore assume that $z_1 = z_2$ for the purpose of this argument, as in

$$Cu^{2+} + Zn = Cu + Zn^{2+},$$

and $\quad \Delta G = -E_{cell}\, zF$

$$= \Delta G_2 - \Delta G_1 = -E_2 zF + E_1 zF.$$

Thus $\quad E_{cell} = E_2 - E_1, \quad (5.28)$

and the "electromotive force" of the cell is equal to the difference between the electrode potentials of the electrodes of which it is made up.

Using the definitions of "anode" and "cathode" in Section 5.3, electrons are consumed at the electrode M_2, which is therefore the *cathode* and is *positively* charged, whereas electrons are produced at the electrode M_1, which is the *negatively* charged *anode*. It should be noted that the electron flow is in the opposite direction to the conventional current flow in the conductor.

The thermodynamic argument used here assumes that all processes are thermodynamically reversible, so that E_{cell} is a measure of the reversible e.m.f. of the cell and assumes that no current flows in the circuit. It is therefore only a measure of the *tendency* of a particular cell to produce an e.m.f.

5.7. The Junction Potential between Two Electrolytes: Symbolic Representation of Cells

In the cell described in the last section, the electrolyte is common to both electrodes—and the metal ions may interfere with one another's movements and with the opposite electrode reactions. To produce a cell possessing an e.m.f. given by (5.28), the electrolytes must be separated into two compartments, connected, say, by some form of "electrolyte bridge" or porous partition (Fig. 5.15).

There will then be a junction between the two electrolytes, and it is found that there is a potential difference set up at this junction, called the *junction potential.*[10] This potential depends on the relative ionic mobilities and on their activities in the electrolyte. The junction potential will affect the e.m.f. of the cell, and in order to minimize its effect with aqueous electrolytes, a "salt bridge" is inserted between the two electrolytes. This consists of an inverted "U"-tube filled with a gel made up of a saturated solution of either potassium chloride

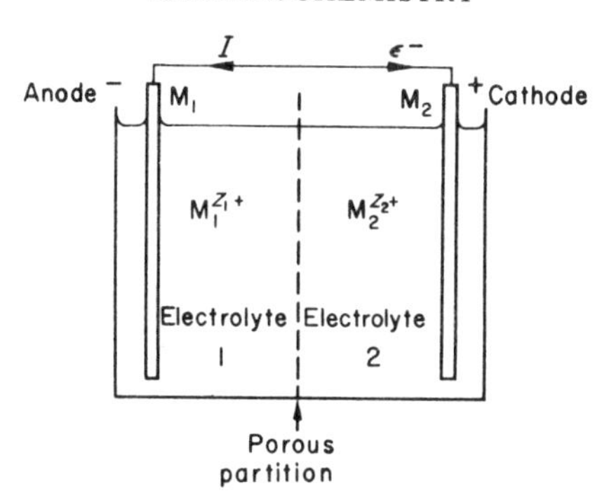

FIG. 5.15. Galvanic cell with separate electrode compartments.

or ammonium nitrate and the gel (e.g. agar agar), (Fig. 5.16). The ionic mobilities of anion and cation in these salts are almost the same, and their concentrations are usually much higher than those of the ions in the electrode compartments.

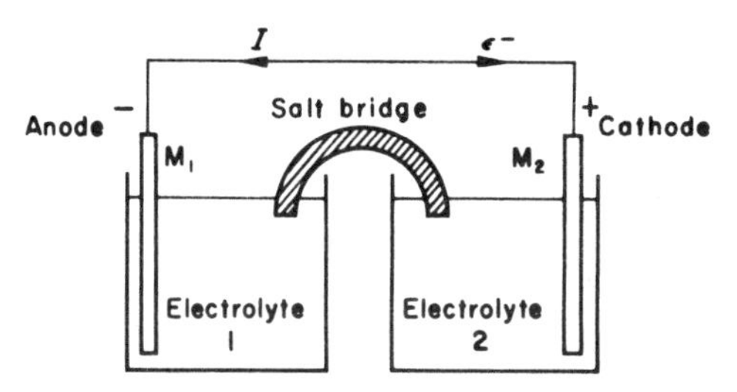

FIG. 5.16. Two electrolytes in a galvanic cell separated by a "salt bridge".

The conventional method of representing a cell of the type discussed in this and the preceding sections is

$$M_1|M_1{}^{z_1+}||M_2{}^{z_2+}|M_2,$$

where the single vertical line represents the junction between the metal electrode and the electrolyte and the double line the junction between the two electrolytes, provided that the junction potential has been made negligible by the use of a salt bridge. If two electrolytes are in contact and no salt bridge is used, a single vertical line is used to separate their symbols. Subscripts may be used to indicate the nature of the electrolyte, e.g. for aqueous electrolytes with no salt bridge,

$$M_1|M_1{}^{z_1+}aq|M_2{}^{z_2+}aq|M_2.$$

An example of a reversible galvanic cell is the Daniell cell,

$$Zn|Zn^{2+}_{aq}||Cu^{2+}_{aq}|Cu$$

which has an e.m.f. of +1·1 V. It should be noted that if the cell is written down in the symbolic manner shown, the cell e.m.f. is given by subtracting the electrode potential of the electrode on the left from the electrode potential of the right-hand electrode. This will give the correct sign to the cell e.m.f.

5.8. The Measurement of Cell e.m.f.'s and Electrode Potentials

A convenient method of measurement of cell e.m.f.'s is by means of a potentiometer in which the e.m.f. of the cell B is compared with that of a "standard cell" C of known e.m.f. (Fig. 5.17). A is a cell of constant e.m.f. (often a lead accumulator) and is connected across the ends of a wire MN of reasonably

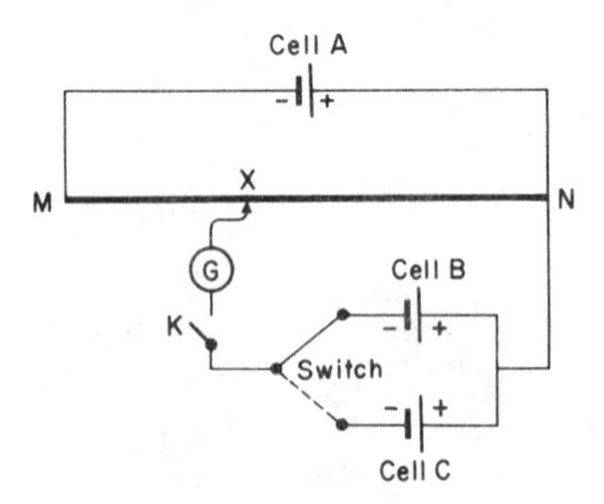

FIG. 5.17. Potentiometer circuit for measurement of cell e.m.f.'s.

high resistance. The cell B is connected so that its e.m.f. is in opposition to that of A, and a sliding contact X is moved until no current flows through the galvanometer G. At this "balance point" the potential drop across XN due to the cell A is exactly balanced by the e.m.f. of B, E_B. Then E_B is proportional to the resistance of XN, which, if the wire MN is of uniform cross-section and material, is proportional to the distance XN_B. The experiment is repeated by use of the switch to bring in the cell C, of standard e.m.f. E_c. The distance XN_C at the balance point is measured and E_C is proportional to XN_C. Then

$$\frac{E_B}{E_C} = \frac{XN_B}{XN_C}.$$

To measure the reversible e.m.f. of cells B and C, no current must pass through them, so that care must be taken to press the galvanometer key K intermittently when the circuit is off balance to reduce the quantity of electricity passed to a minimum.

More sophisticated instruments than this are often used, but all work on a similar principle. Some may be electronic potentiometers, using a high impedance in the circuit to prevent the passage of a significant current through the cell under test, and others a more convenient method of tapping into a resistance XN. Readings to an accuracy of 10^{-4} V are quite easy to attain but other sources of error in applications of metallurgical interest may be so great that the use of more elaborate instruments than this is unnecessary.

The standard cell C must be reproducible, its e.m.f. must have a small dependence on temperature, and it must produce a constant e.m.f. over long periods. The most commonly used is the Weston cell which has the following construction:

$$Cd(12{\cdot}5\%\ \text{amalgam in Hg})\,|\,3CdSO_4, 8H_2O_S\,|\,CdSO_{4(sat)}\,|$$

$$|\,Hg_2SO_{4S}\,|\,Hg.$$

The Weston cell has an e.m.f. of 1·0183 V at 20°C provided very little current is drawn from it.

To measure the *electrode potential* of any particular electrode, the electrode must be made part of a cell. Electrode potentials of single electrodes cannot be measured so that some form of reference electrode must be used, and the formula (5.28) applied. In aqueous solutions the *Standard Hydrogen Electrode* has been chosen as the arbitrary zero of potential against which all other electrode potentials may be measured. In this electrode, pure hydrogen at 1 atm pressure is bubbled through a solution containing hydrogen ions at unit activity. The electronic conductor for the electrode is platinum (Fig. 5.18) whose surface has been made as reactive as possible

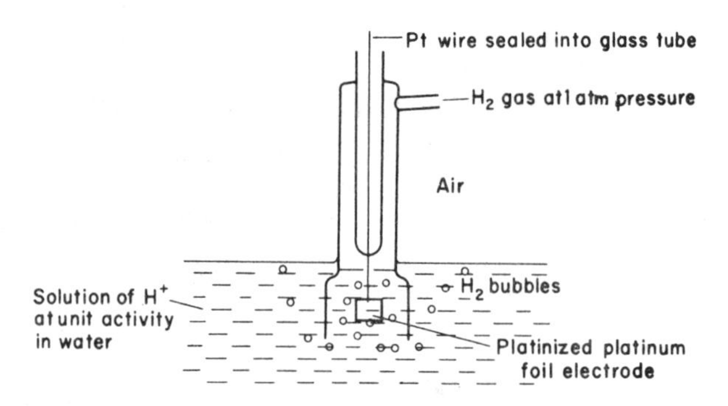

FIG. 5.18. A hydrogen electrode.

by coating with a fine electrolytic deposit of platinum (the so-called "platinized" platinum electrode). The electrode is thus $Pt|H_{2G}(lat)|H^+_{aq}(a_{H^+} = 1)$, and a solution of hydrogen ions containing H^+ at unit activity is 1·19N HCl solution. Here, the electrode potential E_H is E^0_H, the standard hydrogen electrode potential and is given a value of zero volts. If the activity of hydrogen ions in the electrolyte is anything other than unity, using the Nernst equation (5.23) the electrode potential of the hydrogen electrode is then

$$E_H = E_H^0 + \frac{RT}{F} \ln a_{H^+}, \tag{5.29}$$

the electrode reaction being

$$H^+_{aq} + \epsilon^- = \tfrac{1}{2}H_{2G}.$$

At 25°C, E_H^0 is zero and we can rewrite (5.29) as

$$E_H = \frac{RT}{F} \,.\, 2{\cdot}303 \,.\, \log_{10} a_{H^+}$$

$$= -2{\cdot}303 \frac{298R}{F} \text{pH}.$$

Substituting for $\boldsymbol{R}$ (8.31 joules) and $\boldsymbol{F}$ (96,500 coulombs),

$$E_H = -0{\cdot}059 \text{ pH} \quad \text{at } 25°\text{C}. \tag{5.30}$$

Thus the electrode potential of the hydrogen electrode is dependent upon the pH of the electrolyte (pH $= -\log_{10} a_{H^+}$, by definition).

Hydrogen electrodes are inconvenient for practical work, and are replaced by reference electrodes which have known electrode potentials, having been calibrated against a standard hydrogen electrode. For example, an electrode can be constructed by depositing silver chloride electrolytically on a silver wire, and dipping the coated wire into an aqueous solution containing chloride ions:

$$Ag \,|\, AgCl_S \,|\, Cl^-_{aq}.$$

If we set up the cell

$$Pt \,|\, H_{2G} \,|\, H^+(a_{H^+} = 1) \,|\, Cl^-(a_{Cl^-} = 1) \,|\, AgCl_S \,|\, Ag,$$

we find the cell e.m.f. is $+0{\cdot}222$ V at 25°C. The left-hand electrode is the standard hydrogen electrode and has zero potential, so that, using (5.28), and subtracting the left-hand electrode potential from the right-hand electrode potential to obtain the cell e.m.f. (Section 5.6), the electrode potential of the right-hand electrode on the standard hydrogen scale must be +0·222 V. This is the standard electrode potential for the silver/silver chloride electrode as, here, $a_{Cl^-} = 1$. The electrode reaction is

$$AgCl_S + \epsilon^- = Ag_S + Cl^-,$$

so that, from (5.23), and assuming AgCl and Ag to have unit activity (pure solids),

$$E_{Ag/AgCl} = E^o_{Ag/AgCl} - \frac{RT}{F}\ln a_{Cl^-}. \qquad (5.31)$$

We can now use this electrode as a reference electrode to measure an electrode potential, such as that of copper in contact with an aqueous solution of its ions, $Cu \mid Cu^{2+}_{aq}$.

The e.m.f. of the cell

$$Ag \mid AgCl_S \mid Cl^-_{aq} \| Cu^{2+}_{aq} \mid Cu$$

is measured, and, from (5.28)

$$E_{cell} = E_{Cu} - E_{Ag/AgCl},$$

and

$$E_{Cu} = E_{cell} + E_{Ag/AgCl},$$

giving the electrode potential of the copper electrode on the hydrogen scale.

Other convenient reference electrodes include the "calomel" electrode

$$Hg \mid Hg_2Cl_{2S} \mid KCl_{sat} \qquad (E = +0{\cdot}242 \text{ V at } 25°C)$$

and the copper/copper sulphate electrode,

$$\text{Cu} \,|\, \text{Cu SO}_{4\text{sat}} \qquad (E = +0{\cdot}316 \text{ V at } 25°\text{C})$$

which is used in corrosion studies in the field – for example, on pipelines buried in soil.

An electrode used for pH measurements in aqueous solutions is the "glass electrode", which is a silver/silver chloride electrode in HCl solution contained in a tube of glass. The glass is an electrolyte sensitive to hydrogen ions, and, if dipped in an aqueous solution, becomes reversible to hydrogen ions.

$$\text{Ag} \,|\, \text{AgCl}_\text{S} \,|\, \text{HCl}_\text{aq} \,|\, \text{GLASS} \,|\, \text{H}^+_\text{aq}.$$

This electrode's potential, E_G depends on the activity of H^+ in the solution into which it is dipped,

$$E_\text{G} = A - 0{\cdot}059 \text{ pH},$$

where A is a constant. An instrument measuring the electrode potential of the glass electrode against a suitable reference electrode

$$\text{Ag}|\text{AgCl}_\text{S}|\text{HCl}_\text{aq}|\text{GLASS}|\text{H}^+_\text{aq}||\text{KCl}_\text{sat}|\text{Hg}_2\text{Cl}_{2\text{S}}|\text{Hg}$$

can be calibrated in terms of the pH of the test solution and is called a "pH meter". It is used in a wide range of applications (except at very high pH), such as corrosion and electroplating investigations, where pH of a solution is an important variable.

5.9. Reduction and Oxidation Potentials: The Standard Electrode Potential Series

We have seen three types of electrode reaction:

(a) The reduction of a metal ion to cause it to discharge at an electrode

$$\text{M}^{z+} + z\epsilon^- = \text{M}.$$

(b) The reduction of the ion of an element to form the gaseous element at an inert electrode

$$H^{+} + \epsilon^{-} = \tfrac{1}{2}H_{2G} \quad \text{at a platinum electrode.}$$

(c) The reaction involving a metal, a sparingly soluble salt of the metal and the aqueous solution of the anions of the salt

$$AgCl_{S} + \epsilon^{-} = Ag_{S} + Cl^{-}.$$

All these reactions are *reduction* reactions as the species on the left-hand side of the equations are in the "oxidized state" and those on the right-hand side are in the "reduced state". Electrode potentials calculated on the basis of (5.23) are therefore called *reduction potentials.* If the electrode reactions were written down in the reverse direction, the electrode potentials would be those of oxidation reactions, and are called *oxidation potentials.* The sign of E° would be reversed, the logarithm of the reciprocal of the ionic activity would be used, and (5.23) would become

$$E = E^{\circ} - \frac{RT}{zF} \ln a_{M^{z+}}. \tag{5.32}$$

The convention chosen here gives the correct sign convention according to the "European" system and leads to the expression in (5.23) of reduction potentials at electrodes. This system has been adopted by the International Union of Pure and Applied Chemists (I.U.P.A.C.), and is commonly used by electrochemists and corrosion scientists. Unfortunately many American workers and textbooks use the "American" convention, which considers oxidation potentials by (5.23), so that care must be taken before using data from these sources.

A fourth type of reaction can occur at an electrode, the reduction of an ionic species which remains in the electrolyte while the metal of the electrode remains inert. An example of this is a platinum electrode dipping into an aqueous solution of ferrous and ferric ions—two valency states of the same metal,

$$\mathrm{Pt}\,|\,\mathrm{Fe}^{2+}_{\mathrm{aq}},\ \mathrm{Fe}^{3+}_{\mathrm{aq}}.$$

The reaction at the electrode, known as a *"redox" reaction,* is

$$\mathrm{Fe}^{3+}_{\mathrm{aq}} + \epsilon^- = \mathrm{Fe}^{2+}_{\mathrm{aq}},$$

the ferric ion being reduced to the ferrous state.

The *"redox" potential* of this electrode is given by the Nernst equation (5.23),

$$E = E^{\circ} + \frac{\boldsymbol{RT}}{\boldsymbol{F}} \ln \frac{a_{\mathrm{Fe}^{3+}}}{a_{\mathrm{Fe}^{2+}}}, \tag{5.33}$$

which can be generalized as

$$E = E^{\circ} + \frac{\boldsymbol{RT}}{z\boldsymbol{F}} \ln \frac{[\text{activity of reacting species in oxidized state}]}{[\text{activity of reacting species in reduced state}]}. \tag{5.34}$$

In steelmaking, the transfer of sulphur from the molten steel to the molten slag can be considered to be a redox reaction

$$[\mathrm{S}] + 2\epsilon^- = (\mathrm{S}^{2-}),$$

where square brackets indicate a constituent of the metallic phase, round brackets a constituent of the slag phase. The redox potential of the reaction is, using (5.34),

$$E = E^{\circ} + \frac{\boldsymbol{RT}}{2\boldsymbol{F}} \ln \frac{a_{[\mathrm{S}]}}{a_{(\mathrm{S}^{2-})}}.$$

By the use of suitable electrodes, it might be possible to measure the standard potential E°, and determine the potential which would have to be applied to transfer sulphur electrolytically from the steel to the slag under varying activities of sulphur in metal and slag. This electrochemical approach to slag/metal reactions may be fruitful, especially in the study of the kinetics of reactions, and has been discussed by several authors.[11]

If the Standard Electrode Potential E°_{M} is determined for a series of electrode reactions involving metal electrodes in contact with aqueous solutions of their ions, of the type (5.15)

$$M^{z+}_{aq} + z\epsilon^{-} = M,$$

a "Standard Electrode Potential Series" (or "Electrochemical Series") of the type shown in Table 5.1 can be set up. The value of E°_{H}, against which these potentials are measured, is shown for reference. Metals with large positive Standard Electrode Potentials show very little tendency to dissolve in

TABLE 5.1. ELECTROCHEMICAL SERIES (STANDARD ELECTRODE POTENTIALS)

Metal	Electrode reaction	E°_{M} at 25°C (V)
Au	$Au^{3+} + 3\epsilon^{-} = Au$	+1·43
Ag	$Ag^{+} + \epsilon^{-} = Ag$	+0·80
Cu	$Cu^{2+} + 2\epsilon^{-} = Cu$	+0·34
(H	$H^{+} + \epsilon^{-} = \frac{1}{2}H_2$	0)
Pb	$Pb^{2+} + 2\epsilon^{-} = Pb$	−0·13
Sn	$Sn^{2+} + 2\epsilon^{-} = Sn$	−0·14
Ni	$Ni^{2+} + 2\epsilon^{-} = Ni$	−0·25
Cd	$Cd^{2+} + 2\epsilon^{-} = Cd$	−0·40
Fe	$Fe^{2+} + 2\epsilon^{-} = Fe$	−0·44
Cr	$Cr^{3+} + 3\epsilon^{-} = Cr$	−0·74
Zn	$Zn^{2+} + 2\epsilon^{-} = Zn$	−0·76
Ti	$Ti^{2+} + 2\epsilon^{-} = Ti$	−1·63
Al	$Al^{3+} + 3\epsilon^{-} = Al$	−1·66
Mg	$Mg^{2+} + 2\epsilon^{-} = Mg$	−2·40
Na	$Na^{+} + \epsilon^{-} = Na$	−2·71

water—and are known as *noble* metals. *Base* metals have a tendency to dissolve in water or "corrode", shown by their negative Standard Electrode Potentials.

This series gives a rough guide to electrochemical behaviour, but it must be recognized that these are Standard Electrode Potentials, and deviations from standard solutions must be taken into account by means of the Nernst equation (5.23). The temperature may also affect these values, and application of the van't Hoff isochore—or its derivatives—must be used (2.40). Two examples should be sufficient to indicate the use of the Electrochemical Series:

In the cell

$$Fe \,|\, Fe^{2+}_{aq} \,\|\, Cu^{2+}_{aq} \,|\, Cu,$$

$E^{o}_{Fe} = -0{\cdot}44$ V, and $E^{o}_{Cu} = +0{\cdot}34$ V at 25°C. If iron and copper, connected by an electronic conductor, were immersed in standard electrolytes in this cell, then the copper would act as the cathode

$$Cu^{2+} + 2\epsilon^{-} = Cu,$$

iron as the anode

$$Fe = Fe^{2+} + 2\epsilon^{-}.$$

The result would be a tendency for the iron to dissolve in the electrolyte, and this process is known as *galvanic corrosion* of the less noble metal in a system containing two metals. The reversible e.m.f. of the corrosion cell would be

$$E^{o}_{cell} = E^{o}_{Cu} - E^{o}_{Fe} = 0{\cdot}34 - (-0{\cdot}44)$$

$$= 0{\cdot}78 \text{ V},$$

according to (5.28).

Copper is more noble than nickel, so that if an aqueous solution containing Cu^{2+} and Ni^{2+} is passed over nickel powder, the nickel powder will tend to dissolve anodically, and the copper will plate out,

$$Cu^{2+} + Ni = Ni^{2+} + Cu.$$

$$E^{o}_{cell} = E^{o}_{Cu} - E^{o}_{Ni} = 0{\cdot}34 - (-0{\cdot}25)$$

$$= 0{\cdot}59 \text{ V}.$$

This process is known as *cementation* and is used to purify the nickel-bearing electrolyte in the electrolytic refining of nickel. In the same way, scrap iron can be used to remove Cu^{2+} from aqueous solutions and zinc dust to remove gold from aqueous cyanide leach solutions.

Latimer[(12)] can be consulted for data on electrode potentials.

5.10. Concentration Cells

A cell can be constructed of two electrodes of the same metal M dipping into solutions containing ions M^{z+} at different activities a_1 and a_2,

$$M \,|\, M^{z+}(a_1) \,\|\, M^{z+}(a_2) \,|\, M.$$

The electrode potential of the left-hand electrode, by the Nernst equation (5.23), is

$$E_1 = E^{o}_{M} + \frac{RT}{zF} \ln a_1.$$

The electrode potential of the right-hand electrode is

$$E_2 = E^{o}_{M} + \frac{RT}{zF} \ln a_2.$$

Then, by (5.28),

$$E_{\text{cell}} = E_2 - E_1$$

$$= \frac{RT}{zF} \ln \frac{a_2}{a_1}, \tag{5.35}$$

and this depends only on the difference in activities of M^{z+} between the two electrolytes.

The "oxygen electrode" $M \mid O_{2aq}$, a metal in contact with aerated water, has an electrode reaction

$$\tfrac{1}{2}O_2 + H_2O + 2\epsilon^- = 2(OH^-).$$

This is an important reaction, as it tends to be the cathode reaction in corrosion cells set up in aerated aqueous solutions. The Electrode Potential for this electrode is [from (5.34)]

$$E_{O_2} = E^0_{O_2} + \frac{RT}{ZF} \ln \left[\frac{p^{1/2}_{O_2}}{a^2_{(OH)}} \right]$$

assuming that in aqueous solution $a_{H_2O} = 1$. We can now set up a cell $M \mid O_{2aq}(p^A_{O_2}) \parallel O_{2aq}(p^B_{O_2}) \mid M$, whose e.m.f. depends on the relationship (5.35)

$$E_{\text{cell}} = \frac{RT}{ZF} \ln \left(\frac{p^B_{O_2}}{p^A_{O_2}} \right)^{1/2}$$

If $p^A_{O_2}$ is less than $p^B_{O_2}$, A will be the anode, B the cathode in the concentration cell. This is the basis of corrosion due to *differential aeration,* in which a metal structure may be immersed in a large depth of water (Fig. 5.19). The regions close to the surface will be in contact with water of higher oxygen content than those at great depth, and the oxygen-starved metal at great depth will corrode anodically. (The concentration cell will be altered immediately if the metal is iron, because the iron in the anodic region will dissolve whereas the metal

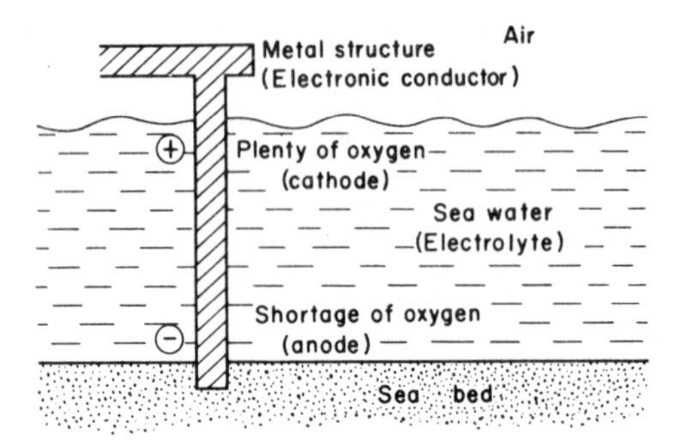

FIG. 5.19. Differential aeration causing a concentration cell to be set up where a metal structure is partly immersed in a large depth of water.

in the cathodic region becomes coated with an oxide layer.) Differential aeration corrosion can be a nuisance because the damage usually occurs at points which are not visible to the casual observer.

5.11. The Use of e.m.f. Measurements with Suitable Cells for the Determination of Thermodynamic Variables

Equation (5.22) shows that e.m.f. measurements can be used to determine free energy changes,

$$\Delta G = -EzF.$$

If the chemical reaction in the cell is suitable, then E will be a measurement of the free energy change accompanying the reaction. The Nernst equation (5.23) for the Electrode Potential of an electrode,

$$E = E^{\circ} + \frac{RT}{zF} \ln a_{M^{z+}},$$

is dependent on the activity of the ion taking part in the electrode reaction—so that if E can be determined experimentally, it might supply a measurement of $a_{M^{z+}}$. We saw in

Section 3.10 that the activity of a component in a solution is related to its partial molar free energy $\bar{G}$ by (3.24)

$$\bar{G}_A - G_A^o = RT \ln a_A,$$

and another thermodynamic variable could be determined. Measurements of E for cells at various temperatures has given their temperature coefficients $(\partial E/\partial T)_p$. Using the Gibbs–Helmholtz equation (2.23),

$$\Delta G = \Delta H + T\left[\frac{\partial(\Delta G)}{\partial T}\right]_p,$$

and substituting for ΔG by (5.22),

$$\Delta G = -EzF = \Delta H - TzF\left(\frac{\partial E}{\partial T}\right)_p.$$

Thus, at temperature T,

$$\Delta H = TzF\left(\frac{\partial E}{\partial T}\right)_p - EzF,$$

and ΔH can be calculated. In the cases in which ΔH has also been determined calorimetrically, results of e.m.f. measurements are in good agreement.[13]

A cell $Ag | Ag^+_{glass} | Ag_xAu_y$ could be set up at a suitable temperature to investigate the reaction

$$Ag + Ag_xAu_y = Ag_{1+x}Au_y,$$

which will give the free energy change when one mole of Ag is added to Ag_xAu_y. If x is large, $(1+x)$ is approximately equal to x, and we have the partial molar free energy of solution of silver in the alloy Ag_xAu_y, $\Delta\bar{G}_{Ag}$.[14] Then

$$\Delta\bar{G}_{Ag} = \bar{G}_{Ag} - G_{Ag}^o = RT \ln a_{Ag}.$$

Wagner[15] discussed the use of cells to determine thermodynamic variables in alloy systems and to determine the activity of components of fused salt mixtures, for example the double cell

$$Pb|PbCl_2, PbBr_{2L}|Graphite, Br_2|PbBr_{2L}|Pb$$

can be used to determine the activity of $PbBr_2$ in $PbBr_2$–$PbCl_2$ melts. The double cell is used to eliminate the liquid junction potential which would be present in the cell

$$Pb\,|\,PbCl_2, PbBr_{2L}\,|\,PbBr_{2L}\,|\,Pb.$$

Solid state electrolytes have been used in cells, such as ZrO_2, MgO in the cell,

$$Ni, NiO\,|\,ZrO_2, MgO\,|\,\text{Oxygen in liquid Pb},$$

to examine the thermodynamic properties of solutions of oxygen in liquid metals, for example in the work of Alcock and Belford.[6]

In e.m.f. measurements with metallurgical systems, care must be taken to ensure that equilibrium has been reached, electrolytes must be pure to avoid secondary reactions, an inert atmosphere must be used, and where alloys are being investigated, the temperature must be high enough to give rapid diffusion rates which prevent a concentration gradient being set up in the electrode.

5.12. The Kinetics of Electrode Processes: Polarization

Our consideration of electrode processes has so far assumed that the electrodes are at equilibrium—that is, if the electrode is part of a cell, no current is passed by the cell. This assumption is necessary for the thermodynamics of the processes, and hence the *tendency* for electrode reactions to occur, to be studied. We have seen (Section 5.8) that it is possible to approximate to thermodynamically reversible conditions to measure reversible cell e.m.f.'s, and hence indirectly measure electrode potentials against a standard electrode whose electrode potential is assumed to be zero.

If the electrode is part of a system in which a finite current is being passed, it is no longer behaving thermodynamically reversibly, and it is found that the electrode potential changes from its reversible value. This change in the electrode potential is known as *polarization*, and is due to several factors. In Fig. 5.12, we see that there is a *free energy of activation* ΔG^* for the electrode process. We know from the treatment of reaction kinetics (Section 4.9) that the rate of the reaction depends on ΔG^* according to (4.26),

$$k = \frac{\boldsymbol{RT}}{\boldsymbol{Nh}} \exp\left(-\frac{\Delta G^*}{\boldsymbol{RT}}\right).$$

$\boldsymbol{RT/Nh}$ is a constant, and the velocity constant k must be related to the flow of current at the electrode so that

$$I = \mathrm{B} \exp\left(-\frac{\Delta G^*}{\boldsymbol{RT}}\right), \tag{5.36}$$

where I is the current at the electrode and B a constant. It is found that if a current is passed at an electrode, ΔG^* will be raised and the electrode potential will alter. Figure 5.20 is an idealized representation of the effects of *activation polarization*, as this form of polarization is called, on the shape of the curve from Fig. 5.12 (shown here as a broken curve). Consider the reaction first as a *cathode reaction*

$$\mathrm{M}^{z+} + z\epsilon^- = \mathrm{M}.$$

ΔG_c^*, the cathode free energy of activation, has increased and

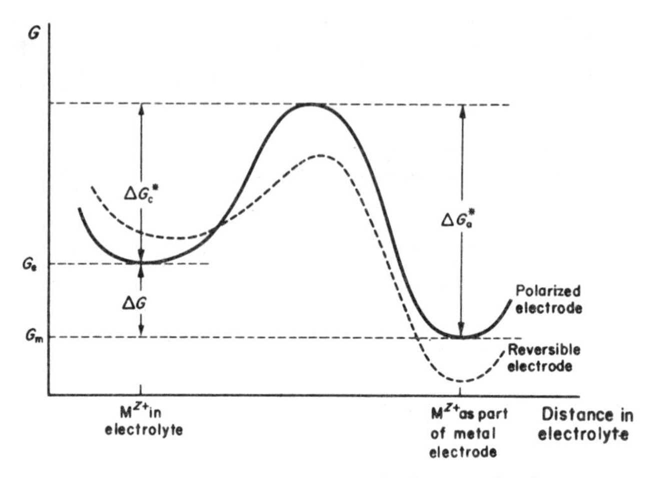

FIG. 5.20. Effect of activation polarization on the free energy change accompanying an electrode reaction (ΔG) and the activation energies of the forward (ΔG_c^*) and reverse reactions (ΔG_a^*). G_e is the free energy of the metal ion M^{z+} in the electrolyte, G_m its free energy when part of the metal electrode.

therefore according to (5.36) the current passed at the electrode will decrease. ΔG, the free energy change accompanying the process, will have a smaller negative value than the reversible value, and therefore, according to (5.22)

$$\Delta G = -EzF,$$

the electrode potential will decrease. If the reaction is now considered as an *anode reaction*,

$$M = M^{z+} + z\epsilon^-,$$

ΔG_a^*, the anode free energy of activation, has also increased, and therefore the current passed at the electrode, if it is an anode, will also decrease according to (5.36). ΔG, the free energy change accompanying the anode process, will have a smaller positive value than the reversible value, and therefore the electrode potential will increase according to (5.22). Activation polarization should decrease with increasing temperature.

It is found that activation polarization is dependent on the current density at the electrode according to the *Tafel equation* (J. Tafel, 1904),

$$\eta^{A} = a \pm b \log i, \tag{5.37}$$

where η^{A} is the *activation overpotential*, i the current density, and a and b are constants depending on the temperature and the electrode; they also depend on the direction of the process, anodic or cathodic. The positive sign in (5.37) applies to the electrode process taking place at an anode, giving the anodic activation overpotential η_{a}^{A}, and the negative sign applies to the cathodic process, giving the cathodic activation overpotential η_{c}^{A}. The *overpotential* at an electrode is the difference between the reversible electrode potential and the polarized electrode potential. Figure 5.21 shows a typical Tafel relationship for a cathode.

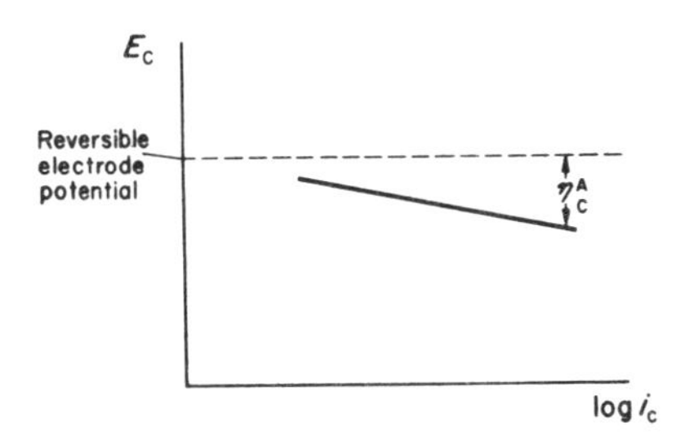

FIG. 5.21. E_c, the electrode potential of a cathode, plotted against log i_c, where i_c is the current density at the cathode. η_c^A is the cathodic activation overpotential which increases as the value of i_c increases.

Activation polarization can be the result of any form of interference with the electrode reaction which impedes one of the reaction steps. An important example of activation polarization is in the discharge of hydrogen gas at a cathode from aqueous solution by the reaction

$$H^+ + \epsilon^- = \tfrac{1}{2}H_{2G}.$$

The *hydrogen overvoltage*, as the activation overpotential η_c^A is usually called in this case, varies according to the metal at which the hydrogen discharges, and appears to be lowered as the catalytic properties of the metal electrode become less marked. For example, η_c^A for hydrogen on a platinum electrode is zero, on iron −0·45 V, and on zinc −1·13 V in a N HCl solution at 20°C. Platinum, iron and zinc show a decreasing tendency to catalytic activity in that order. The discharge of iron and nickel ions from aqueous solution show a high activation overpotential (about −0·6 V), whereas silver discharges with virtually no activation overpotential. (All the values of η_c^A quoted here are for a current density of 1 mA/cm².) The high hydrogen overvoltage on zinc has considerable practical significance because, although zinc is below hydrogen in the electrode potential series, the hydrogen overvoltage is much greater than the zinc activation overpotential and therefore it is possible to plate out zinc cathodically from aqueous solution without serious discharge of hydrogen. Reference 16 should be consulted for a more fundamental treatment of activation polarization.

In addition to activation polarization, irreversible behaviour of electrodes is caused by a change in concentration of the electrolyte at the surface of the electrode, known as *concentration polarization.* When a current is passed by the cell, metal ions may discharge at the cathode too rapidly for more ions to *diffuse* from the bulk of the electrolyte to make up for the ions discharged at the electrode surface. If the activity of M^{z+} ions in the bulk electrolyte is a_b and the activity of M^{z+} ions at the electrode surface is a_s, then the reversible electrode potential E_r, according to the Nernst equation (5.23), will be

$$E_r = E_M^o + \frac{RT}{zF} \ln a_b.$$

The polarized electrode potential E_p will effectively be

$$E_p = E_M^o + \frac{RT}{zF} \ln a_s,$$

and η^C, the concentration overpotential, will be given by

$$\eta^C = E_p - E_r = \frac{RT}{zF} \ln \frac{a_s}{a_b}. \quad (5.38)$$

At the cathode, a_s is less than a_b, so that η^C will be negative and the effective electrode potential will be less than that of the reversible electrode. Conversely, if a metal electrode dissolves anodically, metal ions will build up at the surface of the metal electrode because they are unable to diffuse rapidly enough into the bulk electrolyte. Then a_s will be greater than a_b, η^C will be positive, and the electrode potential will rise as a result of concentration polarization. The higher the current density at the electrode, the greater this effect, and eventually a *limiting current density* will be reached at which the value of a_s will be zero at the cathode, and a maximum (saturation activity) at the anode. These limiting current densities are probably never reached because other ions in solution will take part in the electrode processes before this occurs. Figure 5.22 shows characteristic polarization curves found when this phenomenon is significant. As concentration polarization is diffusion-controlled, it will increase with increasing current density and decrease with increased stirring of a liquid electrolyte and with an increase in temperature.

Another source of polarization is an increase in the electrical resistance of the surface of the electrode. This can be caused by an oxide film deposited on the metal surface or by gas adsorbed on the metal surface. There will be a potential drop

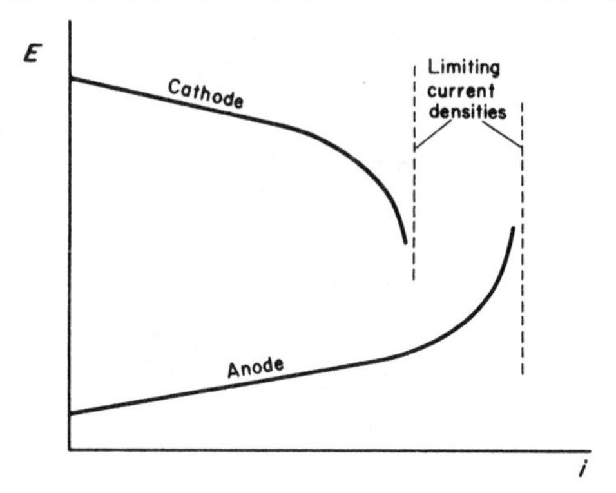

FIG. 5.22. Concentration polarization curves for an anode and a cathode. E is the electrode potential, i the current density.

across such a barrier, and this leads to a *resistance over-potential* η^R.

The total overpotential due to all three types of polarization will be the sum of the three separate overpotentials,

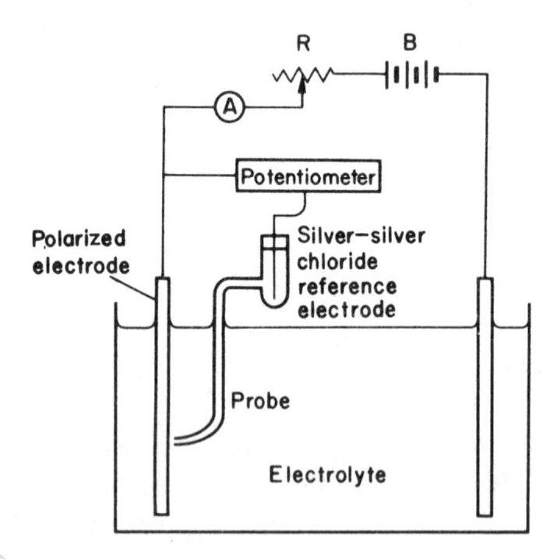

FIG. 5.23. Apparatus for measuring the overpotential at a polarized electrode. A is an ammeter, R a variable resistance, B the external source of e.m.f. supplying the current.

$$\text{Total } \eta = \eta^{A} + \eta^{C} + \eta^{R}, \qquad (5.39)$$

and η^{C} and η^{R} will be more significant at high current densities – such as those used in electrolytic extraction and refining and electroplating of metals. The total overpotential can be measured by placing a "probe" reference electrode close to the surface of the polarized electrode (Fig. 5.23), and comparing the electrode potential thus measured with the known reversible electrode potential. The current density at the electrode can be varied and curves such as those shown in Figs. 5.21 and 5.22 can be produced.

5.13. The Effects of Polarization: Decomposition Voltage: Discharge Potential

The effects of polarization are discussed in Chapter 8, but two effects can be mentioned briefly at this stage. The first is the voltage which must be applied in electrolysis before appreciable reactions take place in an electrolytic cell. If the overall cell reaction is

$$A + BX = AX + B,$$

where A and B are metallic elements, X an acid radical, then ΔG_{cell} will be the free energy change accompanying this reaction. The voltage necessary to cause this reaction to take place reversibly will be given by (5.22)

$$\Delta G_{cell} = -E_{cell} zF.$$

However, both electrodes in the cell will be polarized once a current is passed and also there will be an internal resistance R in the circuit to be overcome.

The voltage necessary to produce a current I in the cell will therefore be

$$V = E_{cell} + |\eta|_{c} + |\eta|_{a} + IR, \qquad (5.40)$$

where η_c and η_a are the cathode and anode overpotentials respectively. Both must be added to E_{cell} irrespective of their sign because both will tend to oppose the applied voltage V. When the current first starts, the cell will be unpolarized and I will be zero, so that the *decomposition voltage*, V_D will then be equal to E_{cell} (Fig. 5.24). However, this point is difficult to establish in practice and measured values of V_D are usually greater than E_{cell} due to polarization.

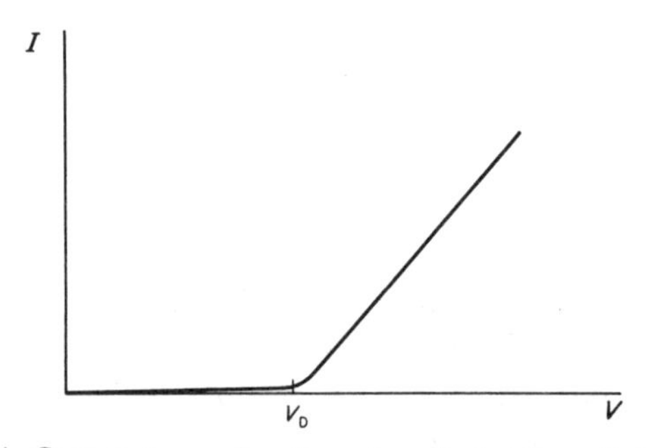

FIG. 5.24. Current I as a function of applied voltage V for electrolysis. V_D is the decomposition voltage.

If the discharge of an ion at an electrode is considered, then the ion will discharge when the potential at that electrode reaches the reversible potential for the particular ion–electrode combination, plus the overpotential for the electrode at the current density used in the process. For a cathodic deposition the discharge potential is $E_r + \eta_c$, and it must be remembered that η_c will have a negative sign. Thus zinc ions will discharge at a zinc electrode when the potential of the electrode is less than $(-0{\cdot}76 - 0{\cdot}20)$ V because $-0{\cdot}76$ V is the reversible electrode potential and $-0{\cdot}20$ V is the activation overpotential for Zn^{2+} (neglecting concentration and resistance polarization). $-0{\cdot}96$ V is the *discharge potential* of the Zn^{2+} ions under these conditions. If the hydrogen overvoltage on zinc is $-1{\cdot}13$ V, E_r for hydrogen zero volt, the discharge potential of the hydrogen ion H^+ will be $-1{\cdot}13$ V. Thus, working under similar conditions, it is possible to discharge zinc from an aqueous solution without discharging hydrogen. Discharge potentials are therefore the potentials at electrodes at which particular ions will discharge, and in an electrolyte containing several metal ions, careful choice of the working cathode potential can allow selective discharge of certain ions, leaving the remainder in solution. This is the basis of the electrolytic methods of extraction and refining of metals, and the polarographic analysis of aqueous solutions.

References

1. For a discussion of the use of the Born–Haber cycle, see GLASSTONE, S. *Textbook of Physical Chemistry*, Macmillan, London, 1953, p. 410.
2. BELL, C. F. and LOTT, K. A. *Modern Approach to Inorganic Chemistry*, Butterworths, London, 1963, Chapter 6.
3. BLOOM, H. and BOCKRIS, J. O'M. *Modern Aspects of Electrochemistry No. 2*. (Ed. J. O'M. BOCKRIS) Butterworths, London, 1959, p. 160.
4. LING YANG and DERGE, G. *Physical Chemistry of Process Metallurgy. Part I*. (Ed. G. R. ST. PIERRE) Interscience, New York, 1961, p. 509.
5. REES, A. L. G. *Chemistry of the Defect Solid State*, Methuen, London, 1954.
6. ALCOCK, C. B. and BELFORD, T. N. *Trans. Faraday Soc.* **60** (1964), 822; and **61** (1965), 443: STEELE, B. C. H. and ALCOCK, C. B. *Trans. Met. Soc. A.I.M.M.E.* **233** (1965), 1359.
7. VERHOEVEN, J. Electrotransport in Metals, *Met. Rev.* **8** (31) (1963), 311.
8. See ref. 1, p. 889.
9. GLASSTONE, S. and LEWIS, D. *Elements of Physical Chemistry*, Macmillan, London, 1961, p. 438.
10. See ref. 1, pp. 931 and 943.
11. WAGNER, C. *The Physical Chemistry of Steelmaking* (Ed. J. F. ELLIOTT), The Technology Press of the Massachusetts Institute of Technology, 1958, p. 245: WARD, R. G. *An Introduction to the Physical Chemistry of Iron and Steel Making*, Edward Arnold, London, 1962, p. 113: LITTLEWOOD, R. *Trans. Met. Soc. A.I.M.M.E.* **233** (1965), 772.
12. LATIMER, W. M. *The Oxidation States of the Elements and their Potentials in Aqueous Solutions*, Prentice-Hall, New York, 2nd edn., 1952.
13. See ref. 9, p. 461.
14. KUBASCHEWSKI, O. and EVANS, E. LL. *Metallurgical Thermochemistry*, Pergamon, London, 1958, pp. 33, 130.
15. WAGNER, C. *Thermodynamics of Alloys*, Addison-Wesley, Reading, Massachusetts, 1952, pp. 97–107, 116ff., 120–5.
16. WEST, J. M. *Electrodeposition and Corrosion Processes*, Van Nostrand, London, 1965, pp. 27–33, 48–52.

CHAPTER 6

Interfacial Phenomena

6.1. Introduction

Most of Chapter 4 was concerned with the chemical kinetics of homogeneous reactions in which the reactants and products are all of the same phase. This approach is necessary to lay a firm foundation for the study of reaction kinetics, but in fact very few reactions of interest to the metallurgist are homogeneous. Reactions between a gaseous and either a liquid or a solid phase, between separate liquid phases or solid phases and between liquids and solids are more important. The reaction involving the transfer of substances between a liquid slag phase and a liquid metal in refining processes such as steelmaking, reduction of a solid oxide by a gas mixture in the blast furnace extraction of metals and the nucleation of gas bubbles in liquid metals are heterogeneous reactions of interest to the extraction metallurgist. All corrosion processes, electroplating and phase transformations in solid alloys are further examples and it is the chemistry of the *interface* between the reacting phases coupled with the transport of reactants and products towards and away from that interface which are important in the kinetics of heterogeneous reactions. Frequently, the temperature of metallurgical processes is so high that the rate of the chemical reaction step in these processes is very fast and is rarely the rate-controlling step in the process, the rate of transport of the reactants and products becoming the dominant factor. This chapter sets out to expand these ideas, first mentioned in Section 4.14, by examining the energy of the interface between phases, nucleation of a new phase in a parent phase, and transport phenomena at interfaces.

6.2. Surface Energy and Surface Tension

A particle in the surface of a liquid or solid has no neighbours on the "free" side – assuming that no matter exists on that side of this "free surface" (Fig. 6.1). The energies of bonds between particles are negative because energy is evolved when a bond is formed. This means that the energy of a particle in a free surface is higher than that of a particle in the bulk phase

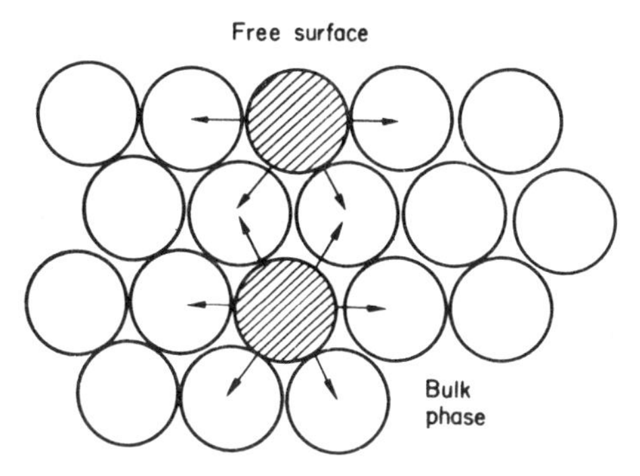

FIG. 6.1. Section normal to free surface of a solid phase. A particle in the free surface is bonded with less neighbours than a particle in the bulk phase.

because it is not bonded to so many neighbours. If the area of a surface is increased, the energy of the surface increases because of this effect, and we can define the *surface free energy* γ as the energy required to create 1 m^2 of new surface, measured in joules per square metre. Liquids clearly possess surface energy, which is a characteristic property of the phase making up the free surface, and because a reduction in the surface area of the liquid results in a reduction in free energy of the system the surface of the liquid will tend to contract as far as possible. Spheres have the smallest surface

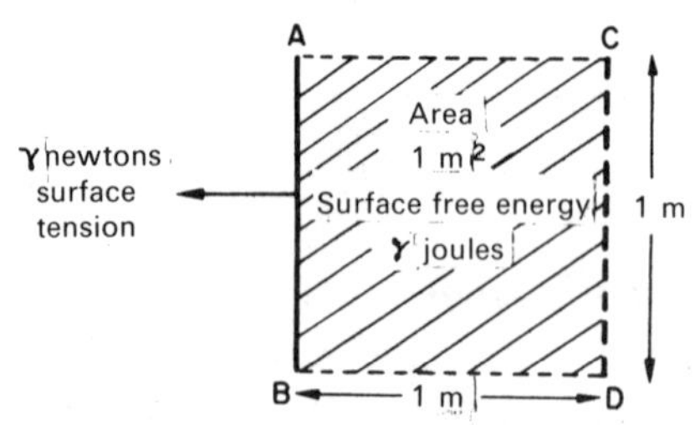

FIG. 6.2. AB is a line 1 m long in the surface of a liquid. The energy required to stretch the liquid surface to CD from AB is γ joules if BD is 1 m and the surface tension is γ newtons per metre.

area to volume ratio of all geometrical shapes, and liquids will tend to form spherical drops as far as other constrictions, such as the force of gravity or frictional forces between the liquid and another phase in motion relative to the liquid, will allow. This tendency of liquid surfaces to contract causes a tensional force in the free surface called the *surface tension*, defined as the force in newtons acting at right angles to a line 1 m long in the surface. The work done in producing $1 m^2$ of new surface is equivalent to the work done in stretching the surface by 1 m along a 1 m line against the surface tension (Fig. 6.2), and this is the surface free energy γ. Thus the surface tension is γ newtons per metre, is numerically equal to the surface free energy and has the same dimensions, $[MLT^{-2}]$. These terms are often used indiscriminately to describe surface properties, and although surface free energy is probably the more fundamental property, their mathematical equivalence justifies the interchange. Because the new surface produced by this stretching is made up of particles brought into the surface from the liquid, γ does not alter as the surface is stretched (unlike elastic tension or energy in a rubber balloon). γ is a characteristic property of the liquid, remaining constant at constant temperature. Solids will certainly have surface free energy, but the concept of surface tension in solids is difficult because they have rigid surfaces which are neither smooth nor able to move to allow the solid to fill the shape of a container – in fact solids act as their own containers. There is some controversy over experiments which have been carried out to measure surface free energy of solids.[1]

Surface tension measurements with liquids involve the measurement of the force required to create surfaces in the form of bubbles or drops or to detach wire rings from the surface of liquids, and reference should be made to other sources for experimental details.[2,3] The surface is never really "free" because it is bounded by a gas and, conventionally, measurements are made in air saturated with the vapour of the liquid. Different values are obtained if the gas is the vapour of the liquid alone or if other gases than air, such as argon, are used. Clearly, surface tension measurements for liquid metals involve special experimental problems, such as oxide contamination of the surface, refractory containers and temperature control, and measurements must be made in an inert gas atmosphere. Some typical surface tension values for liquids are given in Table 6.1, and it will be noted that metals have high surface tensions, a reflection of the strong bonding in highly coordinated metallic structures.

TABLE 6.1. SURFACE TENSION OF CERTAIN LIQUIDS

Liquid	Temperature (°C)	γ (N/m)
Water	100	0·059
Water	20	0·073
Sodium chloride	910	0·106
Steelmaking slag	1600	0·400 (approx.)
Mercury	20	0·480
Zinc	650	0·750
Gold	1130	1·100
Copper	1150	1·100
Steel (0·4%C)	1600	1·560

An example of the importance of surface tension is the problem of the nucleation of gas bubbles in liquid metals. The pressure on a spherical gas bubble in a liquid metal at a certain distance from the surface of the metal can be calculated,

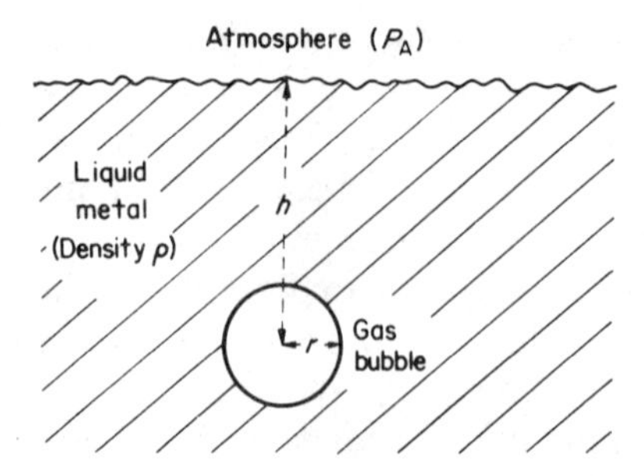

FIG. 6.3. A spherical gas bubble of radius r and in a depth h of liquid metal of density ρ. P_A is the atmospheric pressure above the liquid metal surface.

referring to Fig. 6.3. The total pressure P_B in the gas bubble must, at equilibrium, be equal to the sum of the atmospheric pressure P_A, the pressure due to the head of metal P_M, and the pressure required to maintain the metal/gas surface of the bubble against the surface tension, P_γ. Thus

$$P_B = P_A + P_M + P_\gamma. \tag{6.1}$$

P_M is equal to $h\rho g$, where h is the height of the head of metal above the bubble, ρ is the density of the metal and g is the acceleration due to gravity. P_γ can be calculated as follows: the surface area of a sphere of radius r is $4\pi r^2$. If the radius is increased by a small amount x, the increase in area is

$$4\pi(r+x)^2 - 4\pi r^2 = 4\pi x^2 + 8\pi r x.$$

From the definition of surface free energy at the beginning of this section, the energy required to create this new surface is

$$(4\pi x^2 + 8\pi r x)\gamma.$$

To increase the surface area by this amount, the volume of the sphere must be increased against the pressure P_γ by

$$\tfrac{4}{3}\pi(r+x)^3 - \tfrac{4}{3}\pi r^3 = \tfrac{4}{3}\pi x^3 + 4\pi r^2 x + 4\pi x^2 r.$$

The energy required to produce this increase in volume is

$$(\tfrac{4}{3}\pi x^3 + 4\pi r^2 x + 4\pi x^2 r)P_\gamma,$$

and this must be the same as the energy required to increase the surface area by $(4\pi x^2 + 8\pi r x)$. Thus

$$(4\pi x^2 + 8\pi r x)\gamma = (\tfrac{4}{3}\pi x^3 + 4\pi r^2 x + 4\pi x^2 r)P_\gamma,$$

and if x is very small we can neglect terms in x^2 and x^3 so that

$$P_\gamma = \frac{2\gamma}{r}. \tag{6.2}$$

Thus, from (6.1),

$$P_B = P_A + h\rho g + \frac{2\gamma}{r}. \tag{6.3}$$

If one is trying to remove a gas such as nitrogen present in solution in a liquid metal by lowering the pressure P_A in a vacuum chamber so that, by Sieverts' law (3.7)

$$x'_{[N]} \propto \sqrt{p'_{N_2}},$$

the mole fraction of nitrogen in solution is proportional to the square root of the partial pressure of the nitrogen in the atmosphere, it must be remembered that P_A only applies to the surface of the metal, and P_B, the pressure in a gas bubble, increases as h, the depth of the bubble beneath the metal surface, increases. As r, the radius of the bubble, decreases, P_γ increases, and with small bubbles P_B may be too high for nitrogen to leave the metal and enter the bubble. If a gas bubble is to nucleate within the bulk of liquid metal, by homogeneous nucleation, r will be very small—perhaps of the order of the size of a single molecule. If we consider the possibility of a gas bubble nucleating with a radius of 10^{-9} m at a depth of 0·3 m below the surface of liquid steel, the value of g is

9·81 m/s², the density of liquid steel can be taken as $7{\cdot}4\times 10^3$ kg/m³, and the surface tension of liquid steel (from Table 6.1) as 1·560 N/m. One atmosphere pressure is equivalent to $1{\cdot}01\times 10^5$ N/m², and if $P_A = 1$ atm from (6.3),

$$P_B = 1 + \frac{(0{\cdot}3\times 7{\cdot}4\times 10^3\times 9{\cdot}81)}{1{\cdot}01\times 10^5} + \frac{2\times 1{\cdot}560}{10^{-9}\times 1{\cdot}01\times 10^5}$$

$$= 1 + 0{\cdot}22 + 30{,}900$$

$$\simeq 30{,}900 \text{ atm.}$$

The value of P_γ is so high that the possibility of homogeneous gas bubble nucleation in liquid metals must be ruled out. For a gaseous component to be removed, an existing gas/metal interface must be available—such as the surface of the liquid, gas bubbles deliberately introduced into the metal, vortices formed by a vigorous stirring action, or gas trapped in crevices in the solid surface of the containing vessel. Sims[(4)] discusses the importance of crevices in the hearth as a source of nucleation for carbon monoxide bubbles in the "boil" of an open hearth steelmaking furnace, and Körber and Oelsen[(5)] found that steel covered by a slag became heavily supersaturated with oxygen and carbon in a smooth-walled crucible unless the smooth surface was scratched to allow gas-filled pores in the refractory material to come into contact with the steel. Then a vigorous "boil" of carbon monoxide bubbles was produced by the reaction of carbon and oxygen at the new gas/metal interface

$$[C] + [O] = CO_G.$$

The introduction of bubbles of an inert gas such as argon into liquid steel in a vacuum degassing chamber increases the rate of gas removal by providing a gas/metal interface in the bulk of the melt as well as helping to stir the liquid metal. The fact that air is blown into the liquid metal through the base of a

Bessemer converter must contribute to the rapid removal of carbon in the production of steel in such a vessel by providing a gas/metal interface at which the carbon–oxygen reaction can take place. Rimming steels, which depend on the carbon–oxygen reaction during solidification to produce gas bubbles, are often treated during casting with a volatile compound such as sodium fluoride (boiling point 1700°C) which may assist in nucleation of gas bubbles.

6.3. Interfacial Energy of other than Gas/Liquid Interfaces: The Three-phase Interface

There will be the same tendency to contract at a liquid/liquid interface as there is at a liquid/gas interface, and the *interfacial tension* and *interfacial energy* at such an interface will have the same meaning as the terms discussed in the last section. In fact, *interfacial tension* γ is probably a better name for the general phenomenon at any interface, whether it involves two liquids, two solid phases, a liquid and a solid, a liquid and a gas or a solid and a gas, whilst *surface tension* usually refers to a liquid/air interface.

Kozakévitch[3] discusses methods of determining the interfacial tension between liquid slags and liquid metals, and points out the difficulties encountered in obtaining accurate results. These measurements are of interest in the study of the dispersion of iron droplets in blast furnace and steelmaking slags leading to losses of valuable metal in the slag. It is to be expected that the interfacial tension between two liquids will be less than the higher of the surface tensions of the two liquids because there will be an attraction of the particles of one liquid for those of the other, and this should reduce the tendency of the particles to be pulled inwards, away from the interface – the cause of surface tension at a gas/liquid interface.

According to Stokes' law, which assumes laminar movement of particles in a fluid, large particles should sink or rise (depending on their specific gravity) through a liquid metal faster than small particles. Hartmann[6] found that oxides formed as a result of deoxidation of liquid steel with elements such as silicon and manganese,

$$[Si]+2[O]=(SiO_2),$$

$$[Mn]+[O]=(MnO),$$

floated out of liquid steel at a velocity proportional to their diameters, agreeing with Stokes' law. However, Rosegger[7] found that the rate of removal of deoxidation products depended upon their composition as well as their size. Relatively small alumina-rich particles formed by deoxidation with aluminium,

$$2[Al]+3[O]=(Al_2O_3),$$

separated more rapidly than the larger silicate particles formed by deoxidation with silicon. The small difference in specific gravity could not account for this anomaly, for alumina is more dense than silica, and Rosegger suggests that the rate of removal depends on the interfacial tension between the particles and the liquid steel. A high interfacial tension indicates a small attractive force between the steel and the oxide and hence a small dragging force opposing the motion of the particle. A low interfacial tension should cause a large dragging force and a slower rate of separation. Rosegger states that measurements of the surface tension of oxide mixtures indicate that the addition of alumina increases the surface tension of FeO whilst the addition of SiO_2 and MnO decrease it. Liquid steel is not attracted to alumina, and these facts are used to explain the increased rate of removal of oxides when aluminium is used as a deoxidant. The current interest in the production of "clean" steel should make these suggestions significant and Plöckinger[8] mentions the important effect of interfacial tension on this phenomenon. Clearly aluminium, as well as producing a stable oxide, has advantages over other deoxidants as far as rate of removal of deoxidation products is concerned.

Attempts have been made to relate the energy required to crush mineral particles to the increase in surface free energy as a result of the increased surface area of the particles during comminution. Rittinger's "Law" of Comminution (1867) states that the work done in comminution is proportional to the new surface area produced by the comminution. Experimental results indicate an impossibly high surface free energy for solid minerals, and it is clear that the energy required to deform the minerals and overcome friction has been ignored by Rittinger's hypothesis. It is probable that the amount of energy involved in increasing the surface area is only some 10% of the total energy used, and examples of Rittinger's law being obeyed may be fortuitous. Some workers have suggested that it may apply more accurately to fine grinding where the increase in surface area per unit mass is relatively high.

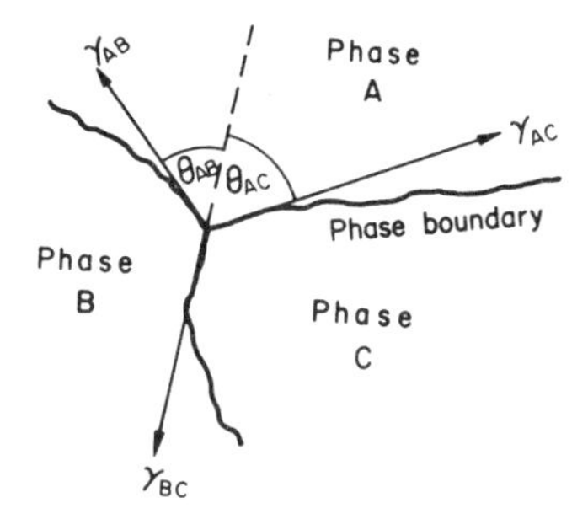

FIG. 6.4. Interfacial tensions at the junction between three phases. γ_{AB} represents the interfacial tension between the phases A and B and is drawn along the tangent to the AB phase boundary at the junction. γ_{AC} and γ_{BC} are drawn similarly.

Where three phases meet at a common boundary, there will be a balance of interfacial tension forces (Fig. 6.4). These can be resolved so that, for equilibrium at the point of intersection,

$$\gamma_{BC} = \gamma_{AB} \cos \theta_{AB} + \gamma_{AC} \cos \theta_{AC}. \qquad (6.4)$$

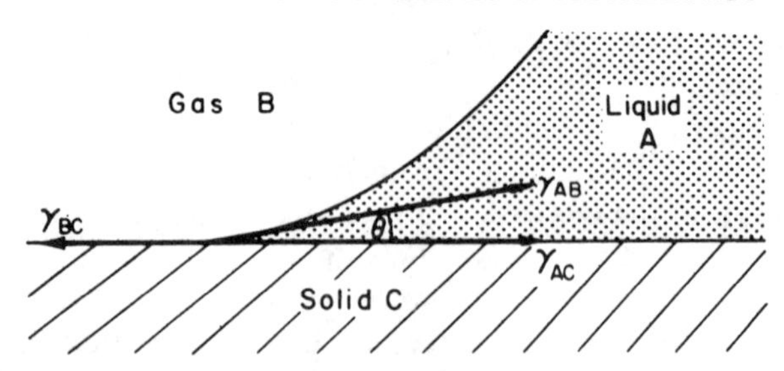

FIG. 6.5. Contact between gas (B), liquid (A) and solid (C), where the solid phase has a plane surface. θ is the contact angle and the liquid "wets" the solid.

γ_{BC}, etc., are the interfacial tensions between adjacent phases and θ_{AB} and θ_{AC} are the angles shown in Fig. 6.4. A special case of this common boundary between three phases is the contact between a liquid, a solid and a gas (Fig. 6.5). If phase A is a liquid, phase B a gas and phase C a plane surface of a solid, θ_{AC} is zero and θ_{AB} is called the *contact angle*, θ. Usually, the gaseous phase B is air, and θ is conventionally measured in the liquid phase. Then, from (6.4),

$$\gamma_{BC} = \gamma_{AB} \cos \theta + \gamma_{AC}. \tag{6.5}$$

If $\theta > 90°$, $\cos \theta$ is negative and $\gamma_{AC} > \gamma_{BC}$. This means that the attraction of A particles for C particles is not as strong as the attraction between A particles, and the liquid does not "wet" the solid (Fig. 6.6). If $\theta < 90°$, $\cos \theta$ is positive and

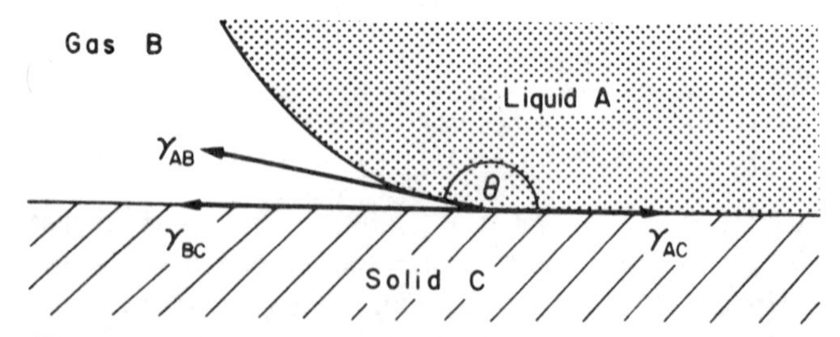

FIG. 6.6. Contact between gas, liquid and solid phases in which the liquid does not "wet" the solid.

$\gamma_{AC} < \gamma_{BC}$. A particles have a relatively strong attraction for C particles and the liquid "wets" the solid (Fig. 6.5). As it would be difficult to measure the actual values of γ_{BC} and γ_{AC}, θ provides the only means of studying their relative strength. When $\theta = 0°$, the liquid will spread completely over the solid surface and when $\theta = 180°$ there will be no attraction between liquid and solid—probably a situation which never exists in nature. Measurements of θ can be made optically, and some examples are given in Table 6.2. Clearly, a polar liquid like water (see Section 5.2) will wet polar solids like glass and ionic solids like sphalerite more easily than non-polar covalent solids like paraffin wax.

TABLE 6.2. SOME EXAMPLES OF CONTACT ANGLES
(Measured in air unless otherwise stated)

Liquid phase/solid phase	θ^0
Water/glass	0
Tin/copper	25
Water/sphalerite(ZnS)	30
Lead/steel (in vacuo)	70
Water/steel	70–90
Water/paraffin wax	110
Mercury/glass	130–150
Mercury/steel	150

An alternative approach to wetting of liquids and solids by liquids was provided by A. Dupré (1869), who defined the *work of adhesion* between a liquid A and a liquid (or a solid) C in the presence of a gas B as

$$W_{AC} = \gamma_{AB} + \gamma_{BC} - \gamma_{AC}. \qquad (6.6)$$

The work of adhesion is the energy required to separate A from C in the presence of the gas B and can be understood if it is remembered that if the area of the interface AC is decreased by 1 m², the areas of the interfaces AB and BC are both increased by 1 m². Then, if the definition of interfacial tension as the energy required to increase the area of the interface by 1 m² is used, the energy required to separate A from C over an area of 1 m² is W_{AC} and is given by (6.6).

In the case of a gas/solid/liquid boundary, (6.5) can be substituted in (6.6) so that

$$\begin{aligned} W_{AC} &= \gamma_{AB} + (\gamma_{BC} - \gamma_{AC}) \\ &= \gamma_{AB} + \gamma_{AB} \cos \theta \\ &= \gamma_{AB}(1 + \cos \theta). \end{aligned} \tag{6.7}$$

Now γ_{AB}, the surface tension of the liquid, and θ can both be measured experimentally so that W_{AC} can be determined. When $\theta = 0°$,

$$W_{AC} = 2\gamma_{AB}, \tag{6.8}$$

and is called the *work of cohesion* of the liquid A because it is the work necessary to pull the liquid apart over an area of 1 cm² with no solid present. Notice that if $\theta < 90°$, $W_{AC} > \gamma_{AB}$ and A wets C. If $\theta > 90°$, $W_{AC} < \gamma_{AB}$ and A does not wet C by our definition of wetting in terms of the contact angle. With complete non-wetting, $\theta = 180°$ and $W_{AC} = 0$.

The *surface roughness* of a solid affects the values of contact angles. Strongly wetting liquids give lower contact angles on rough solid surfaces than on smooth surfaces, but weakly wetting liquids give lower contact angles on smooth surfaces. Bikerman[9] introduces a *roughness factor* ψ into (6.5) to take this into account),

$$\psi(\gamma_{BC} - \gamma_{AC}) = \gamma_{AB} \cos \theta. \tag{6.9}$$

Then (6.7) should read

$$W_{AC} = \frac{\gamma_{AB}}{\psi}(1 + \cos \theta). \tag{6.10}$$

ψ can be calculated by comparing values of $\cos\theta$ for rough and smooth surfaces – where $\psi = 1$ for a perfectly smooth surface – and Bikerman quotes a value of between 1·4 and 2·2 for ψ for ground glass.

The behaviour of metal cast into a mould will depend to some extent on the surface tension of the metal and its ability to wet the mould material. If the mould shape includes a hemispherical portion then, according to (6.2), the pressure head necessary to force the metal into the shape required is dependent on the surface tension of the metal γ and inversely upon the radius r. Differently shaped cavities will require a different expression for the pressure inside the curved surface,[10] and assumptions must be made in the calculation, such as non-wetting of the mould by the liquid metal and no solidification in the mould before the final position of the metal is reached in equilibrium with the pressure head applied. In fact, the metal may wet the mould. In large sand castings, the metal may remelt in contact with the sand and if the metal wets the sand, it may penetrate the cavities between the sand particles, causing a rough surface on the casting. Careful grading of the particles of the sand to reduce porosity and the use of "mould washes" to coat the sand surface with an impermeable barrier (e.g. a zircon wash) will reduce penetration. Reduction of sand porosity will also increase the rate of heat dissipation – thereby increasing the chilling effect of the sand.

The interfacial energy at the boundary between two solid phases in an alloy and between grains of the same metallic phase is of importance in deciding the shape of that boundary.[11] If we consider a non-metallic inclusion or a precipitated intermediate phase (β) in the grain boundary of a metallic structure (α), the shape of the β region will depend upon the interfacial angle (ϕ) between the α and β phases (Fig. 6.7). At equilibrium, the interfacial tension forces must balance and, from Fig. 6.7a,

$$\gamma_{\alpha\alpha} = 2\gamma_{\alpha\beta} \cos\phi. \qquad (6.11)$$

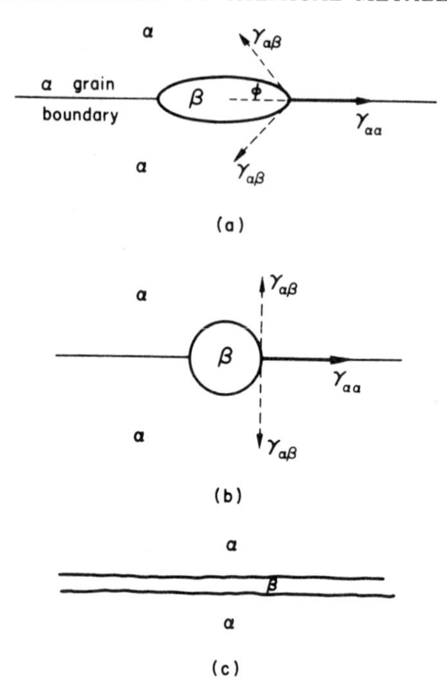

FIG. 6.7. The shape of a precipitate particle β in a grain boundary of α phase, showing the influence of the interfacial angle ϕ on the shape of the β phase. (a) $0 < \phi < 90°$. (b) $\phi = 90°$. (c) $\phi = 0°$.

(This assumes that the β particle is symmetrical with respect to the α grain boundary.) If $\phi = 90°$, the β particle becomes spherical and if $\phi = 0°$, the β phase runs along the grain boundary, forming a film. In this case, if β is a brittle phase, the alloy will tend to fail by intercrystalline fracture. In steels, iron oxide and iron sulphide form in the grain boundary, have a value of ϕ close to zero, and therefore line the grain boundary with a brittle phase which is liquid at hot working temperatures. This is the phenomenon of "hot shortness", prevented by deoxidation of the liquid steel and combination of the sulphur with manganese, whose sulphide does not "wet" the steel at the grain boundary.

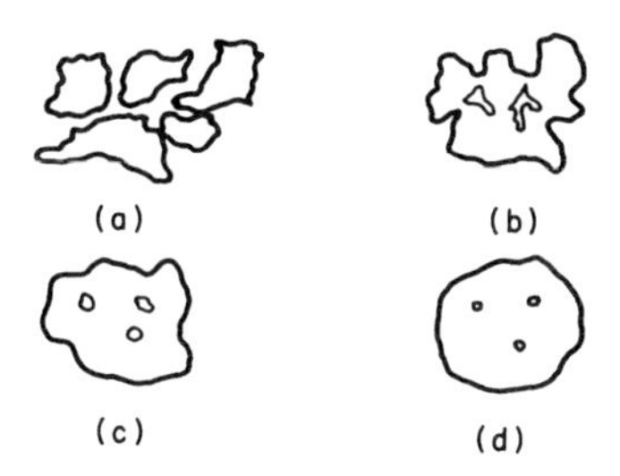

FIG. 6.8. Representation of stages in the sintering of a metallic powder compact. (a) Particles brought together. (b) Formation of pores by sealing off from outer surface. (c) Pores becoming spherical. (d) Almost sound compact.

The shape of grains themselves is controlled by interfacial energy considerations.[11,12] Given the chance, the metallic structure will rearrange itself to reduce its grain boundary energy by increasing its average grain size and producing grains of such a shape that they have a minimum ratio of grain boundary area to volume. This is the phenomenon of grain growth. If possible, a minor phase will take up a spherical shape in the grains of a structure consisting mainly of a major phase. This results in spheroidization of cementite lamellae when a steel structure consisting of a lamellar eutectoid of ferrite and cementite is annealed at about 650°C for long periods.[13]

The sintering of metal and ceramic powders to produce sound parts by heating the powders after compacting under high pressure is dependent on the tendency of a structure to reduce its surface free energy.[14] Atoms or ions diffuse at the points of contact between particles to form a junction; then the pores in the structure become spheroidal and contract by vacancy diffusion (explained in Section 4.13) until a substantially sound part is produced (Fig. 6.8).

6.4. Adsorption

If some substance is present at a higher concentration at the surface of a liquid or a solid phase than in the bulk of that phase, the substance is said to be *adsorbed* on the surface of the phase. Adsorption can be from a vapour or a liquid and is due very often to the fact that substances can form bonds with the phase on which they become adsorbed but that either the bonding is too weak to cause them to enter the bulk phase or their particles are too large to be able to force their way into solution in the adsorbent phase.

The *Gibbs adsorption equation* expresses the fact, first emphasized by J. W. Gibbs (1878), that the concentration of a solute can be higher at the surface of a solution than in the bulk solution:

$$\Gamma = -\frac{c}{\boldsymbol{R}T} \cdot \frac{d\gamma}{dc}, \tag{6.12}$$

where Γ is the excess concentration of solute per square metre of surface (compared with the overall concentration), c is the overall concentration in the bulk solution, $\boldsymbol{R}$ is the gas constant, T the absolute temperature, γ the interfacial energy. The equation applies to dilute solutions, as for more concentrated solutions the concentration term c would be replaced by a, the activity of the solute.[15] This relationship shows that in order that a substance may adsorb at an interface, it must cause a reduction in the interfacial tension at that interface: that is $d\gamma/dc$ must be negative.

I. Langmuir (1916) assumed that gas molecules adsorbed on the surface of a solid in layers one molecule thick (monolayers). At equilibrium, the rate of adsorption will equal the rate of desorption, and if at any time a fraction θ of the adsorption sites are filled, $(1-\theta)$ will be the fraction empty. The rate of adsorption is proportional to p, partial pressure of the gas, and $(1-\theta)$, so that if k is the constant of proportionality,

$$\text{Rate of adsorption} = kp(1-\theta).$$

The rate of desorption will be proportional to θ, so that

$$\text{Rate of desorption} = k'\theta.$$

At equilibrium, these rates are equal, and, at constant temperature,

$$\mathrm{k}p(1-\theta) = \mathrm{k}'\theta,$$

$$\theta = \frac{\mathrm{k}p}{\mathrm{k}p+\mathrm{k}'}, \tag{6.13}$$

which is known as the *Langmuir adsorption isotherm.* In an alternative form, θ is related to the amount of gas adsorbed per unit area of adsorbing surface.[16] Two types of gaseous adsorption can be distinguished, *physical adsorption,* in which the adsorbed molecules are held on the surface by van der Waals bonding, and *chemical adsorption* or *chemisorption* in which the bond strength is much higher—of the order of a true "chemical" bond. The Langmuir isotherm probably applies more strictly to chemisorption as layers more than one molecule thick may be found in physical adsorption. Physical adsorption is dominant at lower temperatures, as it is easily overcome by thermal agitation, whereas chemical adsorption obeys an Arrhenius-type rate equation [see Section 4.6, and equation (4.20)].

$$\text{rate} \propto \exp\left(-\frac{E'}{RT}\right),$$

where E' is the activation energy for the adsorption process. Hydrogen molecules are physically adsorbed on metallic surfaces at temperatures of the order of −200°C, and on raising the temperature they desorb until about −100°C, when chemisorption takes over and becomes stronger as the temperature increases. The hydrogen molecule dissociates and the hydrogen atoms become chemically attached to the metal, whence they can either be in an activated state and take part in a catalysed gas reaction (Section 4.11) or enter the metal by interstitial diffusion (Section 4.12).

Adsorption is used to lower interfacial energy in several instances of importance to the metallurgist. Two substances, such as a mineral oil and water, which are normally completely immiscible and will form two separate liquid layers, can be made to form an *emulsion*—an intimate mixture of small droplets of one liquid dispersed in the other—by adding a substance called an emulsifier which will become adsorbed at the interface, lower the interfacial tension (6.12), and allow small droplets with a large surface area/volume ratio to become stable.[17] A typical emulsifier for use with an oil–water system is a fatty acid salt (soap) or a detergent. These substances consist of a large non-polar group and a polar group (see Section 5.2), e.g. sodium stearate

$$CH_3(CH_2)_{16} \cdot COONa.$$

The large non-polar hydrocarbon chain will dissolve in the oil, and the polar sodium atom will dissolve as a cation in the water, leaving a negatively charged oxygen atom "wetted" by the attraction of the polar water molecules (Section 5.2). The fatty acid anion becomes adsorbed at the oil/water interface and lowers its interfacial tension (Fig. 6.9), so that when a droplet is formed it will remain stable. This type of emulsion is used in the production of cutting fluids for the machining of

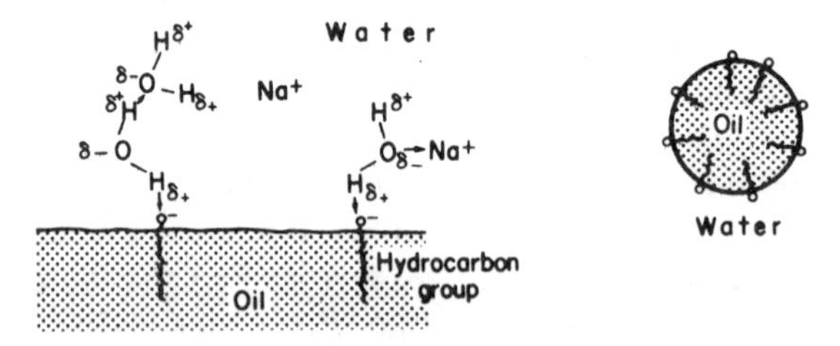

Fig. 6.9. Sodium ion of fatty acid salt dissolves in water, whose polar molecules are attracted to the charged group of the fatty acid anion. The hydrocarbon group is attracted to the oil at an oil/water interface—allowing the formation of small oil droplets dispersed in the water.

metal parts. The water serves to cool the part and tool and the oil to lubricate, although it has been argued that the main function of the oil is to prevent corrosion of the metal—where metal is bared by cutting, the oil and fatty acid adsorb on the surface of the metal and stifle the corrosion process.

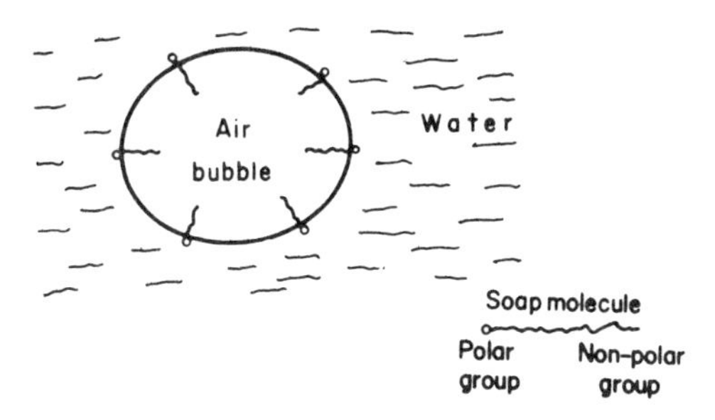

FIG. 6.10. Soap molecules adsorbed at the surface of an air bubble in water.

We saw in Section 6.2 that the excess pressure inside an air bubble in a liquid depends on the interfacial tension γ between the air and the liquid, (6.3). The energy required to form an air bubble will therefore decrease if γ can be lowered. Soaps perform this function in water (Fig. 6.10) by adsorbing at the air/water interface—their polar groups in the water and their non-polar hydrocarbon groups on the air-side of the interface. As in the case of the emulsifier, the adsorption of the soap molecule lowers the interfacial energy of the air/water interface. In the mineral dressing process of *froth flotation*,[18] mineral particles which are not wetted by water become attached to the surface of air bubbles rising through the water—and are thus separated from mineral particles which are wetted by water and consequently sink in the opposite direction (Fig. 6.11); the froth formed by the bubbles runs off the top of the flotation tank or "cell", and when the bubbles break, the floated mineral particles can be recovered by filtration. Small quantities (0·05 kg/tonne of ore in a "pulp" consisting of one-third solids, two-thirds water) of *frothers* are added to lower the energy

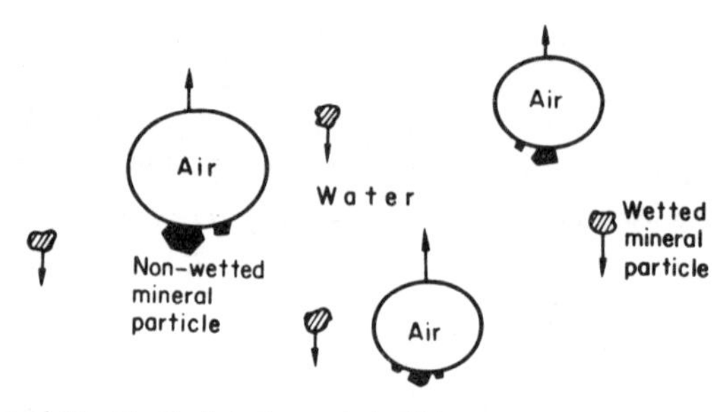

FIG. 6.11. Froth flotation: air bubbles carry non-wetted particles upwards while wetted mineral particles sink.

necessary to produce a bubble and to produce a froth on the surface of the pulp which remains stable until it is scraped off the top of the cell. Frothers are heteropolar substances similar to soaps, which adsorb at the air/water interface and lower the interfacial tension. The frother must produce a froth which is persistent enough to be removed from the cell, but which breaks down almost immediately on leaving the cell – otherwise the plant would be flooded out with froth in a very short time! The lowering of the interfacial energy of the bubbles

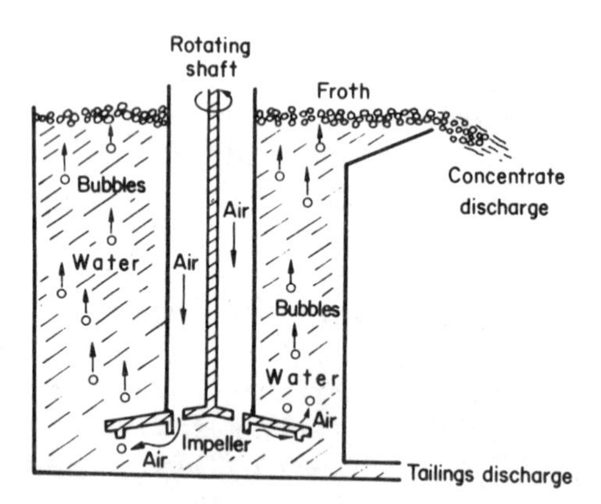

FIG. 6.12. Froth flotation cell: value minerals are usually floated off to form the concentrate; waste ("gangue") minerals leave in the tailings.

can be demonstrated in a cell using paddles to introduce air into the water; if the frother is added to an untreated liquid in the cell, the power requirement of the paddles ("impellers") decreases by about 10%. Figure 6.12 is a representation of a flotation cell. A typical frother is pine oil, based on terpineol, $C_{10}H_{17}OH$: the hydrocarbon group is non-polar and is orientated on the air-side of the interface, whereas the hydroxyl group is polar, and is attracted by hydrogen bonding to the water (Fig. 6.10). Many other frothers are used—alcohols and cresylic acid for example—all having a similar function in lowering the value of γ at the air/water interface.

The "wettability" of the mineral particles is of major importance in flotation, and can be considered on the basis of Fig. 6.5 or 6.6. The boundary between the solid mineral C, water A and air B will be at equilibrium when

$$\gamma_{BC} = \gamma_{AB} \cos \theta + \gamma_{AC}. \quad (6.5)$$

Replacing the symbol C by M for the mineral, A by W for water and B by A for air, in this particular case

$$\gamma_{AM} = \gamma_{WA} \cos \theta + \gamma_{WM}. \quad (6.14)$$

If the mineral is to be "non-wetted" by the water, θ must be greater than 90°. Then $\gamma_{AM} < \gamma_{WM}$. To lower γ_{AM}, a substance which adsorbs at the mineral/air interface must be added. This must not lower γ_{WM} and must therefore present a surface which repels the water. If such a substance—called a *collector*—can be made to adsorb on one mineral rather than on others present, this mineral can be made non-wetted and can be floated selectively from the remaining minerals—and we have a selective concentration process. An example of the use of a collector in froth flotation is potassium ethyl xanthate in the flotation of galena (PbS) from silica (SiO_2). Water will wet both galena and silica, but if a small quantity of the collector is added (25 g/tonne of ore), the contact angle is raised from

0 to 60° on the galena, whereas the contact angle on the silica remains almost zero. The xanthate anion

$$^{-}S \cdot CS \cdot O \cdot C_2H_5$$

is heteropolar, and its polar group ending in a negatively charged sulphur atom is attracted to the lead ions (Pb^{2+}) in the galena. Probably the insoluble lead xanthate can then be considered to exist in the surface of the galena particle. The non-polar ethyl group C_2H_5- is then facing outwards and is less attracted to the water molecules than was the galena—thus, the value of γ_{WM} is raised and as the air/xanthate attraction is higher than the water/xanthate attraction, γ_{AM} is lowered and θ is raised. There appears to be no comparable adsorption

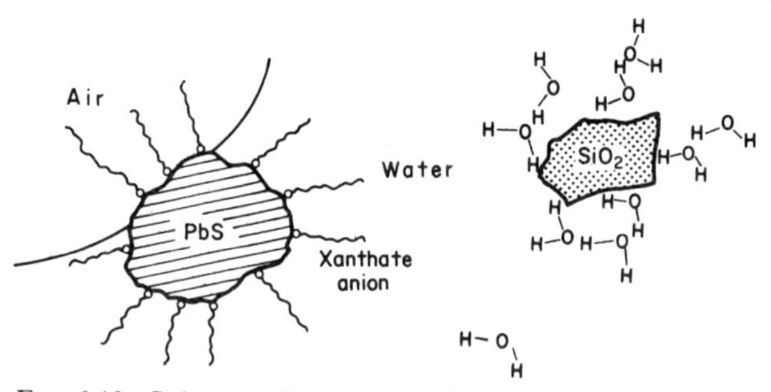

FIG. 6.13. Galena particle with xanthate anions adsorbed on the surface causing it to become attached to an air bubble. The silica particle is wetted by the polar water molecules.

on the silica particles, which remain wetted by the water and sink to leave as tailings, while the galena is floated off as a concentrate (Fig. 6.13). The value of γ_{WA} is important; if this is lowered too much by excessive frother additions, the collector action of the xanthate will be impaired because θ will be lowered to compensate for the increase in γ_{WA}. There is clearly an optimum frother addition above which collector action is nullified. If too much collector is added, the mineral particle is apparently unable to penetrate the air/water interface and consequently will not float. If the length of the carbon chain in the xanthate ion is increased, θ increases – apparently due to the decreasing water/xanthate attraction. Thus, for amyl xanthate anions,

$$^{-}S \cdot CS \cdot O \cdot C_5H_{11},$$

θ is 90° compared with 60° for ethyl xanthate. Measurements of contact angles have been important in the development of flotation, but γ_{AM} and γ_{WM} cannot be measured. The value of θ for a mineral relative to that for another must be higher if the first mineral is to be floated in preference to the second. The maximum value of θ measured for any flotation system appears to be about 110°.

There are many collectors in commercial use – including both cationic and anionic collectors – and their action can be made more selective by the addition of modifying agents, such as CN^- and Cu^{2+}, and by pH control. It has now become possible to float almost any mineral selectively, and the froth flotation process has caused a revolution in mineral dressing and metal extraction processes during this century. It is hoped that this short treatment has been sufficient to demonstrate the importance of interfacial phenomena, including *topochemical reactions*, as reactions on a solid surface are sometimes called, in this process. A comprehensive study can be made by consulting the three books mentioned in ref. 18.

Adsorption plays an important part in the properties of small particles, which have a large surface area/volume ratio – and therefore of *colloids*, which are dispersions of solid or liquid particles in gaseous or liquid media with a particle size intermediate between those of molecules or ions in a true solution and suspensions or emulsions whose dispersed particles may be seen in the optical microscope. Because of their large surface area, the substances adsorbed on their surface largely control the properties – and even the existence – of colloidal particles. Reference 19 can be consulted for discussions of the colloidal state. One property of colloidal dispersions of interest to the metallurgist is the formation of a colloidal dispersion instead of a precipitate in gravimetric analysis. The particles are too small to be retained on a filter pad, and therefore an important precipitate can easily be missed.[20]

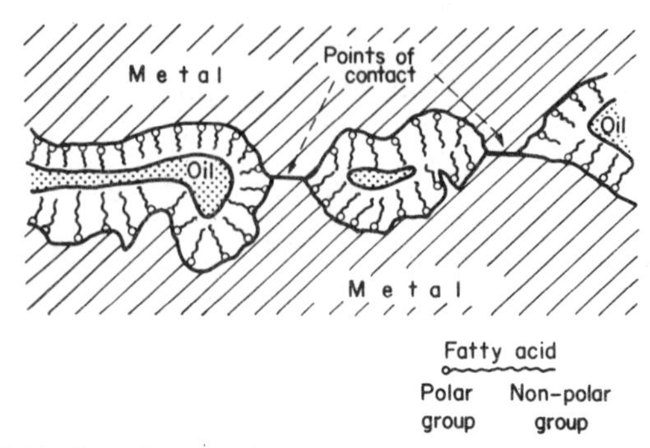

FIG. 6.14. Boundary lubrication of a metal/metal interface: the fatty acid molecules remain adsorbed on the metal surface even where the oil is squeezed out. There is some metal/metal contact.

When two clean metal surfaces are in contact, they become "welded" together by diffusion across the interface. This causes a high frictional force between the two solid surfaces if they move relatively to one another. Severe wear can result in moving parts, and it becomes necessary to use a *lubricant* to separate the metal surfaces. Usually, it is impossible to keep the metallic surfaces completely apart because the thick lubricant breaks down and the metal surfaces then are separated only by adsorbed films of molecular dimensions. This is the condition of *boundary lubrication.* A layer one or two molecules thick could quickly be worn away, so that a lubricant which is quickly adsorbed onto the metal surface is required. All metal/metal contact is not prevented, but it is apparently reduced, and although the coefficient of friction is reduced by a factor of about 10, wear of moving parts can be reduced by a factor of 10^4.[21] Fatty acids, containing a polar group which is strongly adsorbed on the metallic surface, are added to non-polar oils. The non-polar group of the fatty acid presents its surface to the non-polar oil, and the interfacial tension of the oil/metal interface is reduced. The oil may become rubbed away from the metal surface easily, but the fatty acid is much more difficult to dislodge, Fig. 6.14. Reference 22 gives a detailed treatment of friction and lubrication.

6.5. Nucleation

We saw in Section 6.2 the difficulty in nucleating gas bubbles in a liquid metal, requiring a large amount of energy to form the new gas/metal interface. The pressure required to create the new interface was inversely proportional to the radius of curvature of the surface (6.2) and could be very high if the bubble was of molecular dimensions. This problem of *nucleation*—the formation of a new phase separated by a definite boundary from the original phase—exists whenever a phase transformation takes place.[23]

When liquid metals solidify, the problem is the nucleation of a solid particle in a liquid phase, and if the transformation is carried out thermodynamically reversibly—that is always at equilibrium—there will only be one temperature at which the liquid is in equilibrium with the solid, T_0. This temperature is the true "freezing point" of the liquid and is the temperature at which the free energy of the liquid is the same as that of the solid so that

$$-\Delta G_f = G_S - G_L = 0, \qquad (6.15)$$

where G_S is the free energy of the solid metal, G_L the free energy of the liquid metal. ΔG_f is the free energy of fusion of the solid—this must be the same as the free energy change accompanying the solidification process, but of opposite sign. If the temperature is above T_0, the liquid will be more stable than the solid and $(G_S - G_L)$ will be positive. When the liquid is *supercooled* below T_0, $(G_S - G_L)$ will be negative and the solidification process will become spontaneous. If this were the only free energy change taking place, there would be no problem in nucleating the solid particles as soon as the temperature reached or passed below T_o, but in addition to this there will be an increase in the free energy of the system due to the formation of the new interface—this will be the interfacial free energy γ.

G_L is the free energy per mole of liquid metal, so that the free energy per unit volume is

$$\frac{\rho_L}{M} \cdot G_L,$$

where ρ_L is the density of the liquid metal and M is the atomic weight. (If a molecular liquid were solidifying, M would be its molecular weight.) Similarly, the free energy per unit volume of the solid would be (ρ_S/M). G_S where ρ_S is the density of the solid metal. If we consider that the atoms of the liquid metal arrange themselves to form an embryo solid sphere of radius r, the free energy change for nucleation will be

$$\Delta G_N = \left(\frac{\rho_S}{M} \cdot G_S - \frac{\rho_L}{M} \cdot G_L\right) \cdot \frac{4}{3}\pi r^3 + 4\pi r^2 \,.\, \gamma. \qquad (6.16)$$

($\frac{4}{3}\pi r^3$ is the volume and $4\pi r^2$ the surface area of the sphere.) The free energy change associated with the formation of the new phase can be called ΔG_V, the "volume" free energy change, and the free energy change associated with the formation of the nucleus can be rewritten.

$$\Delta G_N = \Delta G_V \,.\, \tfrac{4}{3}\pi r^3 + 4\pi r^2 \,.\, \gamma. \qquad (6.17)$$

But, from (2.18), $\Delta G_V = \Delta H_V - T\Delta S_V$, where $(-\Delta H_V)$ is the latent heat of fusion per unit volume $(L_{f,V})$ and $(-\Delta S_V)$ is the entropy of fusion per unit volume which, from (2.12), is given by

$$\Delta S_V = \frac{-L_{f,V}}{T_o}. \qquad (6.18)$$

Now if T_t is the temperature at which the embryo forms, $\Delta T = T_o - T_t$, the degree of supercooling, and, from (6.17) and (6.18)

$$\Delta G_V = -L_{f,V} - \frac{T_t(-L_{f,V})}{T_o}$$

$$= -\frac{\Delta T \, . \, L_{f,V}}{T_o} \tag{6.19}$$

Thus, from (6.17),

$$\Delta G_N = -\frac{\Delta T \, . \, L_{f,V}}{T_o} \, . \, \frac{4}{3}\pi r^3 + 4\pi r^2 \, . \, \gamma. \tag{6.20}$$

Differentiating (6.20) with respect to r,

$$\frac{d(\Delta G_N)}{dr} = -\frac{\Delta T \, . \, L_{f,V}}{T_o} \, . \, 4\pi r^2 + 8\pi r \, . \, \gamma,$$

and ΔG_N will have a maximum value when

$$\frac{-\Delta T \, . \, L_{f,V}}{T_o} \, . \, 4\pi r^2 + 8\pi r \, . \, \gamma = 0$$

or

$$r_c = \frac{2\gamma \, . \, T_o}{\Delta T \, . \, L_{f,V}}. \tag{6.21}$$

In order that the nucleation process becomes spontaneous, ΔG_N must be negative and therefore a *critical radius* r_c can be postulated at which ΔG_N reaches a maximum. Only if the embryo forms so that its radius is greater than r_c will the volume free energy change become more important than the surface free energy change, and then the embryo can continue to exist as a true nucleus on which more solid phase can grow. This argument can be illustrated by a diagram such as Fig. 6.15, showing the growing importance of the ΔG_V term as r increases. It will be noted from (6.21) that r_c is dependent on ΔT, the degree of supercooling. Even if the shape of the embryo is not spherical, r_c will still be inversely dependent on ΔT. As ΔT increases, r_c decreases and the statistical probability of an embryo of at least radius r_c appearing increases – and therefore the rate of nucleation should increase.

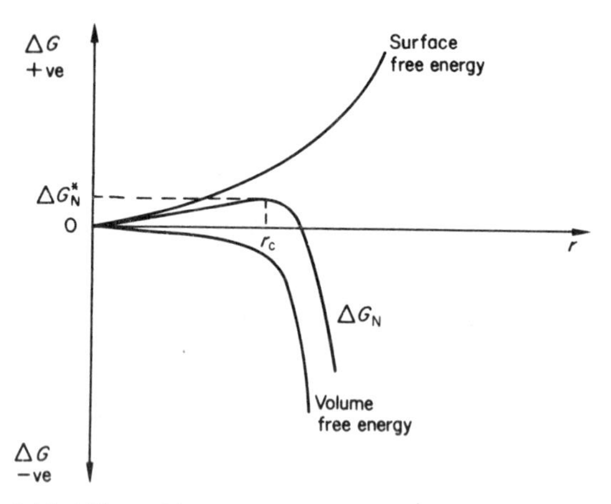

FIG. 6.15. Effect of increasing the radius of a particle, r, on the free energy of nucleation of the particle, ΔG_N, at a constant degree of supercooling, ΔT. ΔG_N^* is the free energy of activation for nucleation.

The statistical probability of finding of radius r_c can be represented by a Boltzmann factor in a similar argument to that employed for the collision theory of reaction kinetics in Section 4.7, so that

$$\text{Rate of nucleation} \propto \exp\left(-\frac{\Delta G_N^*}{RT_t}\right), \qquad (6.22)$$

where ΔG_N^* is the free energy of activation for nucleation and is the minimum increase in free energy required by an embryo to form a nucleus of radius r_c (Fig. 6.15). It will be seen that not only does r_c decrease as ΔT increases, but ΔG_N^* decreases, causing an increase in the rate of nucleation (Fig. 6.16). As ΔT decreases, and eventually is zero—that is $T_t = T_0$ and the system reaches equilibrium, r_c [from (6.21)] will become infinite, and the rate of nucleation will be zero. This is the

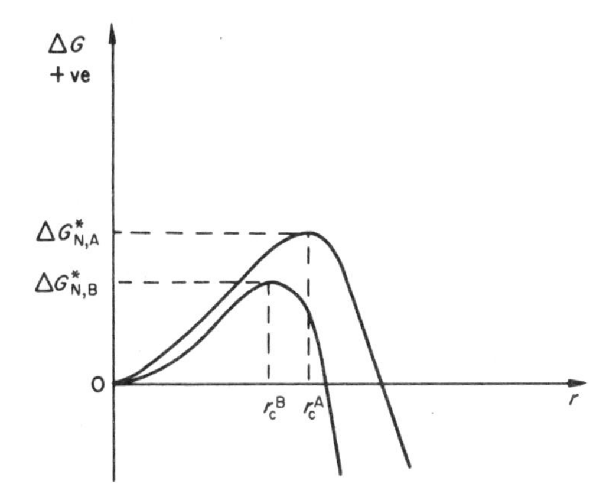

FIG. 6.16. Effect of increasing degree of supercooling from ΔT_A to ΔT_B on the critical radius of an embryo, r_c, and the free energy of activation for nucleation, ΔG_N^*.

reason for supercooling on solidification—in which the temperature of the system drops below T_0 on cooling until nucleation occurs—then the latent heat of fusion released by the liquid raises the temperature to T_0 until solidification is complete (Fig. 6.17). The faster the cooling rate, the greater the degree of supercooling, the faster the rate of nucleation and consequently the finer the grain size of the solidified metal.

Hollomon and Turnbull[(23)] point out that the necessary value of ΔT to achieve a finite rate of nucleation is about one hundred times that usually found in practice, so that the value of this theoretical approach is only qualitative. The reason for this anomaly is the assumption that nucleation is *homo-*

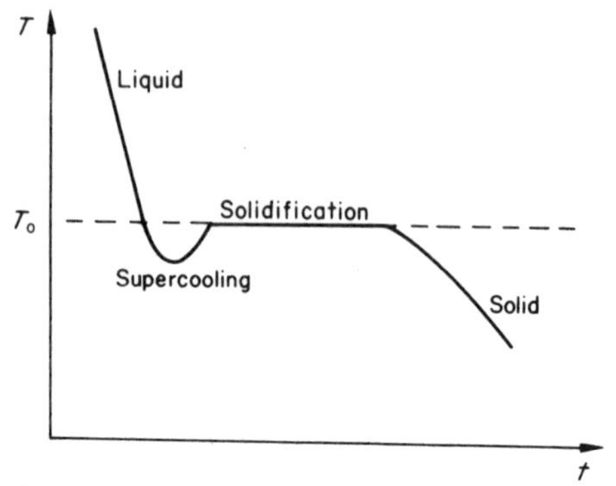

FIG. 6.17. Graph showing a fall in temperature T of a metal with time t as the metal is cooled slowly through the freezing point T_0.

geneous, that is taking place in the bulk liquid with no third phase present, and no solid nuclei of the new phase existing in the melt. Just as with the gas bubble nucleation discussed in Section 6.2, some form of solid/metal interface must exist in the melt for the liquid/solid transformation to take place. The nucleation of a phase on existing interfaces is called *heterogeneous* nucleation.

No quantitative theory of heterogeneous nucleation of general application has been produced, but some important qualitative indications are reasonably well established. If a metal is to solidify on a third phase, which might be solid impurities in the metal—for example, oxides formed as a result of deoxidation practice—or the walls of the container, it is reasonable to suppose that there must be a net attraction between the solid metal and the third, nucleating phase—that is the solid metal must *wet* the solid surface of the third phase more strongly than the liquid wets the third phase. The treatment of Fig. 6.5 and Fig. 6.6 in Section 6.3 can now be applied, where instead of a gas phase B is the liquid metal, A the solid metal, and C the solid nucleating phase. Again, from (6.5),

$$\gamma_{BC} = \gamma_{AB} \cos \theta + \gamma_{AC}$$

at equilibrium. For good wetting of the nucleating phase C by the liquid metal, γ_{BC} should be reasonably low, and for the solid metal to nucleate easily on the phase C, there must be a lower value of γ_{AC} than γ_{BC}, indicating a similarity in structure between A and C, and a strong attraction of A for C. If γ_{AC} is very low, θ must be low to achieve a balance. Once the solid metal A has nucleated on C, we now have effectively a nucleus with radius r greater than the critical radius r_c, and growth of the nucleus can proceed. If a fine-grained solid structure is required, a large number of nuclei can be provided by the addition of "grain-refiners" to the liquid metal. These substances either are insoluble in the metal and form a suitable nucleating surface or react with the metal to form the nucleating phase. Titanium carbide has been added to aluminium alloys for this purpose, and this substance has a similar crystal structure, lattice parameters and bonding to the aluminium, producing low values of γ_{AC} and γ_{BC} in (6.5), and lowering the amount of supercooling ΔT necessary to produce a rapid rate of nucleation.

The solidification of eutectic alloys can be approached from the point of view of the interfacial energy between phases forming simultaneously from the melt and between these phases and the liquid alloy. The mode of growth may be "modified" if some element added to the melt in small quantities is adsorbed on the surface of nuclei in such a way that the interfacial energy is altered. Examples of this phenomenon may be the modification of the aluminium–silicon eutectic by additions of sodium to the melt[24] and the modification of the grey cast iron eutectic of austenite and graphite to form a spherical graphite eutectic by additions of magnesium or cerium to the melt.[25] Further reading on the solidification of metals is suggested in ref. 26.

The theory of nucleation in the solidification of metals can be extended to transformations in solid alloys. If precipitation of β phase in α is to take place by cooling the alloy X from the temperature indicated in Fig. 6.18, T_0 is the equilibrium transformation temperature. Homogeneous nucleation of the β phase can be expressed by (6.20) modified to include $L_{t,V}$, the latent heat of transformation of β to α per unit volume,

$$\Delta G_N = -\frac{\Delta T \,.\, L_{t,V}}{T_o} \cdot \frac{4}{3}\pi r^3 + 4\pi r^2 \,.\, \gamma. \qquad (6.23)$$

γ is the interfacial free energy between the α and β phases.

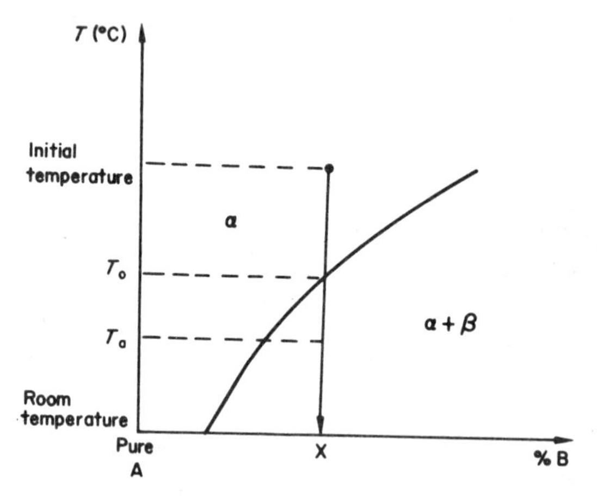

FIG. 6.18. Part of a phase diagram for the binary alloy system A–B. The alloy X is shown cooling from an initial temperature in the α phase region to room temperature in the $(\alpha+\beta)$ phase region. T_a is the ageing temperature used in precipitation hardening.

The critical radius r_c for nucleation of β is given by a similar expression to (6.21),

$$r_c = \frac{2\gamma \,.\, T_o}{\Delta T \,.\, L_{t,V}}. \qquad (6.24)$$

The rate of nucleation will be inversely dependent on ΔT, and will obey the relationship (6.22)

$$\text{Rate of nucleation} \propto \exp\left(\frac{\Delta G_N^*}{RT_t}\right).$$

As in solidification, a higher degree of supercooling ΔT will give a greater number of nuclei and therefore a greater number of β particles in the final microstructure. However, in order that the embryo may grow, atoms of A and B must be transferred across the interface by diffusion, and this process is controlled by an activation process expressed by an equation such as (4.31), so that

$$\text{Rate of growth of embryo} \propto \exp\left(-\frac{E_D}{RT_t}\right), \tag{6.25}$$

where E_D is the activation energy for diffusion in the structure.

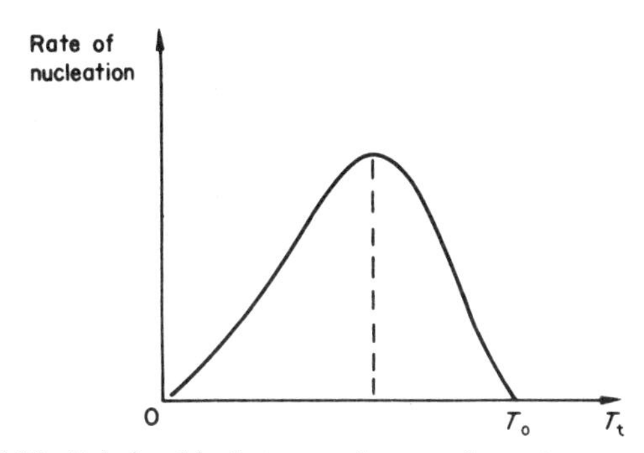

FIG. 6.19. Relationship between the transformation temperature and the rate of nucleation in a diffusion-controlled process.

Although ΔG_N^* decreases as T_t decreases, the rate of diffusion will decrease by (6.25), and eventually (6.25) becomes more important than (6.22), the rate of nucleation going through a maximum (see Fig. 6.19). If supercooling is so great that T_t is room temperature (e.g. as a result of quenching the alloy), no β phase will precipitate and the structure will consist of wholly α phase at room temperature.[27]

Heterogeneous nucleation may occur in solid state transformations and nucleation sites may be provided by dislocations—occurring in grain boundaries, twin boundaries, slip planes and sub-grain boundaries for example—or by non-metallic inclusions such as oxides and nitrides.[23]

If the degree of supercooling is too small for an effective rate of nucleation and the rate of diffusion is low, an intermediate precipitate β' will often form, whose structure is more closely matched to that of the parent α phase than is that of the β phase.[28] The coherency between the α and the β' phases is higher than that between the α and β phases, and a hardening of the alloy is the result; this is known as *precipitation hardening*, and is produced by quenching α to room temperature and "ageing" for a definite period at some intermediate temperature T_a, below T_0. The reason for the increased rate of nucleation—which incidentally produces a very widely dispersed precipitate and consequently increased hardening effects—is the lowering of γ, the interfacial energy between the α and the nucleating phase. r_c and ΔG_N^* will be lowered in consequence and precipitation becomes more likely. If the structure is "overaged", the metastable β' phase will transform by diffusion to the stable β phase as $G_\beta < G_{\beta'}$, and the resultant loss of coherency between the precipitate and the matrix causes a softening of the alloy. The precipitation hardening of aluminium—4% copper alloys has been studied extensively, and the precipitation sequence incorporates two intermediate structures, θ'' and θ', before the equilibrium θ phase ($CuAl_2$) finally appears.[29]

$$\alpha \rightarrow \alpha + \theta'' \rightarrow \alpha + \theta' \rightarrow \alpha + \theta.$$

α is a face-centred cubic solution of copper in aluminium, θ'' is tetragonal with similar lattice parameters to the α matrix in two directions but dissimilar in the third. θ' has a more pronounced tetragonal structure than θ'', and in θ this structure has lost almost all its coherency with the matrix.

Metals can be precipitated from aqueous solution by reduction with hydrogen under high pressure at temperatures of the order of 120°C following a leaching operation.[30] The leaching operation and purification of the solution can also be carried out under high pressure and at elevated temperatures,[31] and the combination of the two processes of leaching and precipitation has provided one of the most important developments in extraction metallurgy in recent years. In the precipitation of cobalt metal from aqueous solution by hydrogen gas

$$Co^{2+}_{aq} + H_{2_G} = 2H^{+}_{aq} + Co_S,$$

cobalt ions will only be reduced at an existing solid surface—thus nucleation must be caused by the introduction of solid nuclei. These may be cobalt particles, or can be non-metallic particles, such as graphite. Stainless steel and pyrex glass also promote nucleation. Nickel reduction has also been studied, and again nuclei must be provided for precipitation. Nickel powder and zirconia have been used—particles of zirconia present in the centre of nickel powder particles offering the possibility of producing dispersion-hardened alloys by powder metallurgy, with the hardening phase "built in" to the powder. The addition of certain organic substances can affect the rate and mode of precipitation of the nucleus. Anthraquinone accelerates the rate of nickel precipitation by lowering the activation energy for the process—apparently by increasing the "effective surface" of the solid particle for further precipitation; some form of adsorption may be involved, but the action seems to be highly specific—it does not increase the rate of cobalt precipitation—and is so far imperfectly understood.[30] Certain substances such as anthraquinone can cause cobalt to precipitate preferentially on certain crystal faces, and this produces particles with a predominance of certain faces—if these faces are better catalytic surfaces than others in the crystal (Section 4.11), the cobalt powder can be rendered highly catalytic for certain industrial processes such as high temperature organic synthesis. Again adsorption may be the basis of the mechanism of controlled precipitation, just as it is in the action of glues in the production of smooth electrolytic deposits (Section 8.6).

6.6. Mass Transport in Heterogeneous Reactions

If a reaction is taking place involving more than one phase, the rate of reaction may depend on several steps—for example, at the boundary surface between a liquid metal and a liquid slag in a smelting process, a reaction involving transfer of sulphur from the metal into the slag will have three steps:

(a) Transport of the reactants to the slag/metal interface. Oxygen ions, (O^{2-}), in the slag and sulphur, [S], dissolved in the metal will be the reactants in this case.

(b) Chemical reaction at the interface, represented by the equation

$$(O^{2-}) + [S] = (S^{2-}) + [0].$$

(c) Transport of the products away from the slag/metal interface—i.e. sulphide ions, (S^{2-}), in the slag, and oxygen, [O], dissolved in the metal.

The chemical reaction in step (b) will obey the laws of chemical kinetics as they were discussed in Chapter 4, and will be temperature-dependent according to the Arrhenius equation (4.20)

$$k = A \exp\left(-\frac{E}{RT}\right).$$

In many metallurgical processes, temperatures are high, and the rate of this step will be high and unlikely to be rate-controlling. Only if E, the activation energy for the reaction, is high will the rate of (b) become as important as the mass transport in (a) and (c) in controlling the overall rate of the heterogeneous reaction. In this particular case—of sulphur transfer from metal to slag—the activation energy of the chemical reaction (b) is of the order of 100 kcal/mole in iron-making,[32] which is high (see Section 4.6) and is likely to mean that (b) is rate-controlling.

It will be apparent that the rate of transport of substances in phases and at the boundary between phases, is of considerable importance in the majority of metallurgical processes, and the remainder of this chapter will be concerned with transport phenomena. If the phase in which transport is taking place is a solid, only one transport process is operative—the diffusion of components in the solid state, which was discussed in Sections 4.12 and 4.13, Fick's law of diffusion may be applied,

$$J = -D\frac{\mathrm{d}c}{\mathrm{d}x}, \tag{4.28}$$

and

$$\frac{\partial c}{\partial t} = \frac{\partial}{\partial x}\left(D\frac{\partial c}{\partial x}\right), \tag{4.29}$$

where J is the mass of diffusing substance passing in unit time through unit area of a plane at right angles to the direction of diffusion—the "flux". $\mathrm{d}c/\mathrm{d}x$ is the concentration gradient across the plane and D is the diffusion coefficient. Solid state diffusion is slow and likely to be rate-controlling.

If the phase is a liquid or a gas, bulk movement of the phase may occur as a result of natural convection or stirring (forced convection) and this mode of transport is likely to be more important than molecular diffusion. Darken[33] considers the two cases of streamline and turbulent flow in fluids. Streamline, laminar or viscous flow is smooth, and if the flow is in the x direction,

$$J_x = -D\frac{\mathrm{d}c}{\mathrm{d}x} + v_x c, \tag{6.26}$$

where v_x is the component of fluid velocity in the x direction. The equation expresses the combined contribution of diffusion

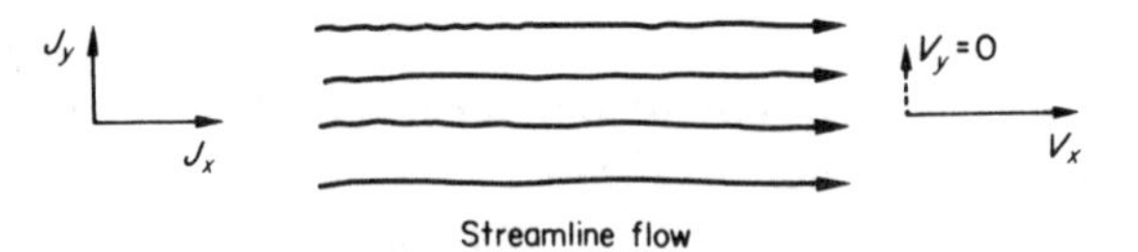

Streamline flow

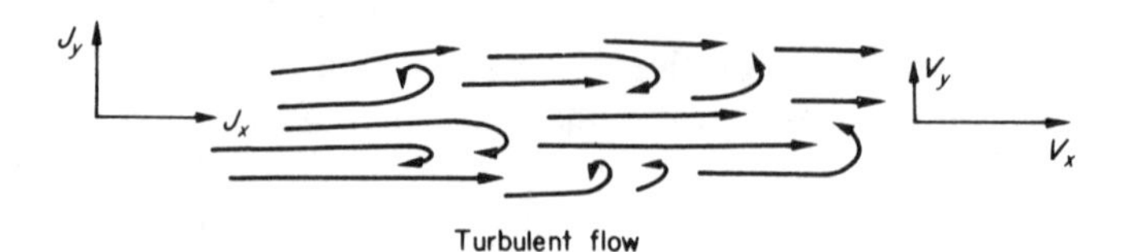

Turbulent flow

FIG. 6.20. Streamline and turbulent flow in a fluid. v_x and v_y are the components of mean bulk velocity in the x and y directions respectively, J_x and J_y the corresponding rates of mass transfer per unit area.

and bulk movement to mass transport in such a case. Transport in the direction at right angles to the flow direction will be by diffusion alone because streamline flow is characterized by the absence of bulk movement in the y direction, so that, by (4.29)

$$J_y = -D\frac{\mathrm{d}c}{\mathrm{d}y}. \tag{6.27}$$

If the velocity of flow is increased, eventually the smooth streamline flow begins to break up into "eddies", causing dispersion of the fluid in the y direction (Fig. 6.20). This results in random mass transport in the y direction, and at any point, the velocity and direction of flow will fluctuate in a random manner, even though the mean velocity and direction of flow of the fluid remain constant.[34] The eddies are large blocks of fluid and these become mixed perpendicular and parallel to the main direction of flow so that a new mode of transport called *eddy diffusion*, which is more rapid than molecular diffusion, is added to the two modes in streamline flow. Thus

$$J_x = -D\frac{\mathrm{d}c}{\mathrm{d}x} + v_x c - D_e\frac{\mathrm{d}c}{\mathrm{d}x}, \tag{6.28}$$

where D_e is the coefficient of eddy diffusion. It would be difficult to distinguish between the contributions to transport made by the three modes, but in a well-stirred fluid, it might be reasonable to suppose that most of the transport is due to eddy diffusion. Szekely[35] studied the rate of mixing in an open hearth steelmaking furnace by introducing a radioactive isotope of gold as a tracer into the centre door of the furnace and taking metal samples at intervals through two side doors. He found that a uniform distribution of the gold was obtained in 3–11 min and, assuming that transport was by eddy diffusion, a value of 200 cm^2/sec for D_e could be calculated. This can be compared with values of the order of 10^{-5} cm^2/sec for molecular diffusion in liquid iron at 1550°C.[36]

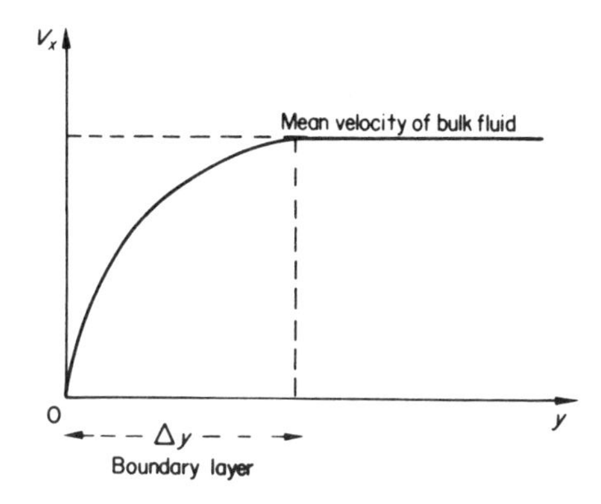

FIG. 6.21. Velocity of fluid flow (v_x) in the x direction, parallel to the surface of a solid, as a function of y, the distance in the fluid from the solid/fluid interface. At a distance Δy the velocity reaches that of the bulk fluid phase.

The mass transfer conditions occurring near the boundary of a fluid phase will be dependent on the nature of the phase on the other side of the boundary. If we first consider transport in a fluid phase at the boundary between the fluid phase and a solid phase, the fluid velocity parallel to the interface will decrease as the distance from the boundary decreases, due to friction between the solid and fluid which is transmitted to the bulk phase by succeeding layers of fluid (Fig. 6.21). At a distance Δy from the interface, the velocity of flow parallel to the interface will reach v_x, the velocity of flow of the bulk fluid in the x direction, and we can postulate the existence of a *boundary layer* of thickness Δy in which the velocity v_x gradually decreases from the bulk velocity to zero at the solid/fluid interface.

If a substance is being transferred from the solid to the fluid – an example being the *leaching* of ores and concentrates in which minerals in the solid surface are taken into aqueous solution – the concentration of the substance being transferred will vary across the boundary layer from c_0 at the interface to c_b in the bulk fluid. The thickness Δy of the boundary layer is difficult to calculate, but a convenient approach is to postulate an *effective boundary layer*, which is a stagnant layer in the fluid at the boundary across which mass transport is only by molecular diffusion because the layer is not disturbed by convection. Whether the flow in the bulk fluid is streamline or turbulent, because close to the interface turbulent flow conditions probably disappear, the postulation of the stagnant

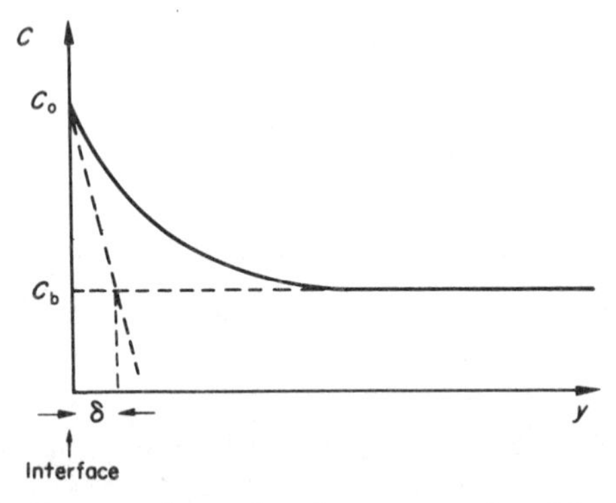

FIG. 6.22. Concentration c of a substance transferred from a solid to a fluid, plotted against y, the distance from the solid/fluid interface. c_0 is the concentration in the fluid at the interface, c_b the concentration in the bulk fluid. δ is the effective boundary layer thickness.

static boundary layer is valid and its thickness will depend on the viscosity of the fluid, the substance being transferred and the amount of convection in the fluid—the greater the stirring, the thinner the boundary layer. Figure 6.22 shows how the *effective boundary layer thickness* δ can be defined by drawing a tangent to the $c-y$ curve at the interface.

δ is the distance between the interface and the point of intersection of the tangent and the horizontal through c_b, the concentration in the bulk fluid, drawn on the graph of c against y. From (6.27), if the transport in the direction y is only by diffusion across the static boundary layer,

$$J=-D\left[\frac{dc}{dy}\right]_{y=0},$$

where $[dc/dy]_{y=0}$ is the concentration gradient at the interface. From Fig. 6.22,

$$\frac{c_o-c_b}{\delta}=-\left[\frac{dc}{dy}\right]_{y=0},$$

and therefore

$$J=D\frac{c_o-c_b}{\delta}. \tag{6.29}$$

If $J=(dm/dt)/A$, where dm/dt is the rate of mass transfer and A is the area of the interface,

$$\frac{dm}{dt}=\left(\frac{DA}{\delta}\right)\cdot(c_o-c_b), \tag{6.30}$$

or

$$\frac{dm}{dt}=k_m(c_o-c_b), \tag{6.31}$$

where k_m is the *mass transfer coefficient.*

This approach to mass transfer at a boundary is similar to that for heat transfer between a well-stirred fluid and a solid. Effective boundary layer thicknesses δ are difficult to estimate, and usually determination of k_m is as far as it is possible to extend this treatment. Taking a value of δ of 0·003 cm, Darken[37] has calculated rates of carbon removal from steel in an open hearth furnace which agree reasonably well with observed rates. However, δ will vary with D, the diffusivity of the substance being transported, and is therefore not an independent variable. Experimental confirmations of theoretical treatments of this type are very rare, and at best the treatment provides only a semi-quantitative approach to the problem of reaction rates at an interface.

The static boundary layer theory is developed for transfer in a fluid at the interface between the fluid and a solid. It depends on the rigidity of the solid to reduce the velocity of flow of the fluid to zero at the interface. If the boundary phase is a liquid, the surface is no longer rigid, and will move in sympathy with the movement of the fluid. Only if the liquid has a much higher viscosity than the fluid can this treatment be valid for transport in the fluid – for example at a gas/liquid interface, or at a liquid metal/slag interface – where the first-named phase has the lower viscosity and is the phase in which transport is being considered. (The viscosity of liquid metals is of the order of 10^{-2} poise, that of liquid slags from 0·2 to 20 poise, depending on the slag composition and temperature, and that of gases is of the order of 10^{-5}.)

The problem of transport in the more viscous phase – such as the liquid metal in a metal/gas system – is not the same, because no static boundary layer can be possible if there is very little external restraint on the surface. Machlin[38] considers the case of an inductively – stirred melt, and derives an expression for the mass transfer across a boundary layer which was rigid but not static. This expression can be represented by (6.31), and the mass transfer coefficient k_m contains D, the diffusion coefficient, and v, the velocity of the liquid metal relative to the surface. v depends on the frequency of the electrical supply and on the electrical power used by the induction coil, and cannot be measured or calculated accurately. Once again, investigation leads to the mass transfer coefficient, and no further, so that whether a static boundary layer or a rigid boundary layer are assumed, it is not possible to differentiate between the results of experiments on this basis.

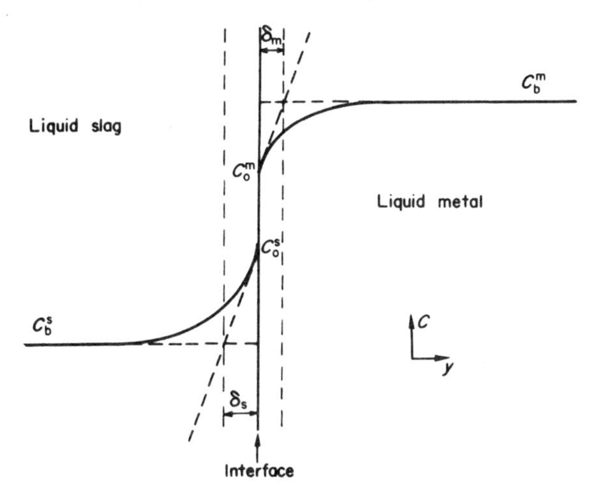

FIG. 6.23. Transfer in liquid slag and metal phases across the slag/metal interface. δ_m is the effective boundary layer thickness in the metal, δ_s the effective boundary layer thickness in the slag.

When a substance is transferred from one fluid phase to another, the mass transport on both sides of the boundary layer becomes important. Whitman[39] proposes a *two-film theory* in which a static boundary layer exists at both sides of the interface. Figure 6.23 represents transfer of a substance from liquid metal to liquid slag, and, using (6.31) for the transport to the interface in the metal,

$$\frac{\mathrm{d}m}{\mathrm{d}t} = k_m^{\mathrm{m}}(c_{\mathrm{o}}^{\mathrm{m}} - c_{\mathrm{b}}^{\mathrm{m}}).$$

If the substance does not accumulate at the interface, this rate of mass transfer must be the same as that away from the interface in the slag,

$$\frac{dm}{dt} = k_m^s(c_o^s - c_b^s),$$

where c_o^m and c_b^m are the concentrations at the interface and in the bulk of the metal phase, and c_o^s and c_b^s apply similarly to the slag phase. If these two rates are equal,

$$k_m^m(c_o^m - c_b^m) = -k_m^s(c_o^s - c_b^s),$$

with the change in sign to take account of the different directions of mass transfer relative to the interface. Thus the ratio of the mass transfer coefficients in metal and slag is given by

$$\frac{k_m^m}{k_m^s} = \frac{c_b^s - c_o^s}{c_o^m - c_b^m}. \qquad (6.32)$$

D_s, the diffusivity in the slag phase, will differ from D_m, the diffusivity in the metal—and δ_s and δ_m, the effective boundary layer thicknesses in the two phases, will also differ. This is not surprising if we consider the original example of sulphur transfer

$$[S] + (O^{2-}) = (S^{2-}) + [O].$$

Diffusion of atomic sulphur in the metal can hardly be the same phenomenon as diffusion of ionic sulphur in the ionic slag structure.

Clearly, from (6.30), the importance of reducing δ by increased stirring and increasing A by increasing the area of the slag/metal interface in slag/metal reactions cannot be over-emphasized. One important method of achieving both aims is by introducing gas bubbles into the liquid metal. This is the basis of all steelmaking processes using a basic slag (rich in lime) to absorb sulphur and phosphorus from the steel; this reaction only takes place at the slag/metal interface. The carbon monoxide bubbles produced by the reaction of carbon and oxygen in the steel

$$[C]+[O]=CO_G$$

will rise to the surface. In the Bessemer converter, air is introduced at the base of the vessel, and the gas bubbles (containing nitrogen, and carbon monoxide as a result of the above reaction) rise to the surface, but in the open hearth furnace, the turbulence is due to the reaction alone.

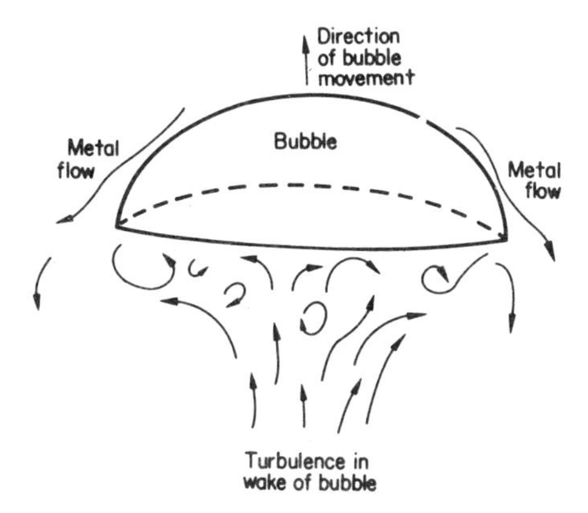

FIG. 6.24. Impression of a "spherical cap" gas bubble rising through a liquid metal, causing stirring of the metal.

Richardson[40] discusses the probable form of bubbles of this type and suggests that when they reach the slag/metal interface, they are about 5 cm across and of "spherical cap" rather than spherical shape (Fig. 6.24). He discusses the evidence available for this suggestion, and points out that the liquid immediately behind the bubble will be drawn up with the bubble, forming a turbulent wake. When the bubble passes into the slag phase, a skin of metal will be carried with it – this eventually falls off and runs back into the metal phase. Bubbles bursting at a liquid/gas interface will also spray liquid droplets into the gas phase. The importance of the evolution of such bubbles in stirring the reacting phases (reducing δ) and increasing A, the area of the interface, must be clear.

6.7. Evaporation

The evaporation of substances from a liquid/gas interface can be one of the steps in a reaction process—and with the growth of interest in the vacuum purification of metals (vacuum melting and degassing) this step has been the subject of some interest.

The mass of gas molecules, J, striking 1 cm^2 of a liquid surface per second can be calculated[41] from the kinetic theory of gases,

$$J = p_i(M/2\pi \boldsymbol{R}T)^{1/2}, \qquad (6.33)$$

where p_i is the vapour pressure of the molecules of the component i in the liquid phase, $\boldsymbol{R}$ is the gas constant, and M is the molecular weight of i. If the system is at equilibrium, the rate of evaporation is equal to the rate of condensation—and if a fraction α of the molecules striking the surface condense on the surface, the maximum rate of evaporation $(dm/dt)_{max}$ is given by

$$\left(\frac{dm}{dt_{max}}\right) = \alpha \, . \, A \, . \, J = \alpha p_i(M/2\pi \boldsymbol{R}T)^{1/2} \, . \, A. \qquad (6.34)$$

This is under conditions of "molecular evaporation", in which the rate of evaporation is so small that the mean free path of the vapour molecules exceeds the distance between evaporating and condensing surfaces, and $(dm/dt)_{max}$ is only achieved if all the molecules evaporating are subsequently condensed—neither returning to the surface nor establishing a concentration gradient in the vapour phase. A perfect vacuum in contact with the liquid is assumed.

From (3.3a),

$$p_i = p_i^o \, . \, a_i,$$

where p_i^o is the vapour pressure of pure i, and a_i is the Raoultian activity of i in the liquid phase. Now, from (3.4),

$$a_i = \gamma_i \, . \, x_i,$$

where γ_i is the Raoultian activity coefficient of i, and x_i the mole fraction of i in the liquid phase. Thus, from (6.34),

$$\left(\frac{\mathrm{d}m}{\mathrm{d}t}\right)_{\max} = \alpha \, . \, \gamma_i \, . \, x_i \, . \, p_i^o (M/2\pi \boldsymbol{R} T)^{1/2} \, . \, A, \qquad (6.35)$$

which is known as the *Langmuir equation* (I. Langmuir, 1913).

The partial pressure of i in the "vacuum" is not zero, so that the net rate of evaporation, $\mathrm{d}m/\mathrm{d}t$, is given by

$$\frac{\mathrm{d}m}{\mathrm{d}t} = p_i^o \, . \, \gamma_i (M/2\pi \boldsymbol{R} T)^{1/2} \, . \, \alpha(x_i - x_i^o) \, . \, A, \qquad (6.36)$$

where x_i^o is the mole fraction of i in the liquid which would be in equilibrium with the partial pressure of i already present in the atmosphere. $\mathrm{d}m/\mathrm{d}t$ is the rate of mass transfer of i into the vacuum, and (6.36) can be written

$$\frac{\mathrm{d}m}{\mathrm{d}t} = k_m(x_i - x_i^o), \qquad (6.37)$$

where k_m is the mass transfer coefficient.

Sehgal and Mitchell[42] studied the kinetics of sulphur transfer at low pressures from an inductively stirred liquid steel bath in a vacuum melting unit, and from the value of the mass transfer coefficient concluded that the rate of evaporation of the sulphur from the metal surface was the rate-controlling step rather than mass transport in the metal (using the Static and Machlin Boundary Layer theories discussed earlier in this section). Ward[43] carried out a similar analysis of manganese loss from vacuum induction-melted steel, and concluded that the rate controlling step varied between transport in the metal boundary layer, evaporation from the metal surface and transport in the gas boundary layer as the pressure in the unit increased. These papers[42,43] should be consulted as examples of the diagnostic approach to reaction kinetics in heterogeneous reactions.

There is a strong possibility that surface active agents can influence transfer between metals and gases by blocking the interface and influence movement in the boundary layer. Bradshaw and Richardson[36] discuss the possibility of the adsorption of oxygen from the metal phase onto the gas/metal interface accounting for the slow rate of transfer of nitrogen from liquid steel in vacuum degassing. Small amounts (0·05 wt.%) of oxygen are known[44] to lower the surface tension of liquid iron by as much as 30% so that strong adsorption is indicated (see the discussion of the Gibbs adsorption equation (6.12) in Section 6.4). Kozakévitch and Urbain[44] found that increase in the oxygen content of liquid iron reduced the rate of nitrogen transfer into the metal from nitrogen gas bubbles by a large factor. The nitrogen content of vacuum degassed steel is usually about 20 times the equilibrium content—and this surface phenomenon may be important in contributing to the apparent slow transfer rate.

Thus, as in the final example of vacuum degassing and associated processes, it is hoped that the importance of interfacial phenomena in almost every field of metallurgy will have been made apparent by this chapter, which by its nature can only serve as a brief introduction to a large number of topics associated with these phenomena. Thermodynamics provides a good signpost to the possibilities of a reaction or process, but reaction kinetics is the map which unfortunately can only give a vague idea of the conditions of the route in our present state of knowledge.

References

1. BIKERMAN, J. J. *Surface Chemistry*, Academic Press, New York, 2nd edn., 1958, p. 177.
2. Ref. 1, pp. 4–23: KOZAKÉVITCH, P. Measurement of the Surface Tension of Liquid Metals. Paper 1E of National Physical Laboratory Symposium No. 9, *The Physical Chemistry of Metallic Solutions and Intermetallic Compounds*, H.M.S.O., London, 1959.
3. HUGHEL, T. J. (Ed.) *Liquids: Structure, Properties and Interactions*, Elsevier, Amsterdam, 1965. Contribution by KOZAKÉVITCH, P., p. 243, Surface Tension of Liquid Metals and Oxide Melts.
4. SIMS, C. E. *Trans. A.I.M.M.E.* **172** (1947), 176.
5. KÖRBER, F. and OELSEN, W. *Mitt. Kaiser-Wilhelm-Inst. Eisenforsch*, **17** (1935), 39.
6. HARTMANN, F. *Stahl und Eisen*, **65** (1945), 216.
7. ROSEGGER, R. *Radex-Rundschau*, No. 6, 1958 and translated in abridged form in *Iron and Coal Trades Rev.*, July 15, 1960, p. 131.
8. PLÖCKINGER, E. *J.I.S.I.* **201** (1963), 576.
9. Ref. 1, p. 352.
10. Ref. 3, Contribution by HARVEY, D. J., p. 285. The Importance of the Surface Tension of Metals in Some Engineering Problems.
11. SMITH, C. S. *Trans. A.I.M.M.E.* **175** (1948), 15.
12. SMITH, C. S. *Met. Rev. (Inst. of Metals)*, **9**, No. 33 (1964), 1.
13. United States Steel Corp. *The Making, Shaping and Treating of Steel*, 1957, p. 789.
14. KINGERY, W. D. *Introduction to Ceramics*, Wiley, New York, 1960, p. 369.
15. For a derivation of the Gibbs Adsorption Equation, see DAVIES, J. T. and RIDEAL, E. K. *Interfacial Phenomena*, Academic Press, New York, 1961, p. 196.
16. GLASSTONE, S. *Textbook of Physical Chemistry*, Macmillan, London, 1953, p. 1198.
17. Ref. 1, p. 155; ref. 15, p. 359.
18. SUTHERLAND, K. L. and WARK, I. W. *Principles of Flotation*, Australasian Inst. Mining and Met., Melbourne, 2nd edn., 1955: GAUDIN, A. M. *Flotation*, McGraw-Hill, New York, 1957: KLASSEN, V. I. and MOKROUSOV, V. A. *An Introduction to the Theory of Flotation*, Butterworths, London, 1963.
19. Ref. 16, p. 1231 ff.; ref. 15, p. 343 ff., p. 386 ff.
20. WILSON, C. L. and WILSON, D. W. (Eds.) *Comprehensive Analytical Chemistry*, Vol. IA, Elsevier, Amsterdam, 1959, p. 448.
21. Ref. 15, p. 434.

22. BOWDEN, F. P. and TABOR, D. *The Friction and Lubrication of Solids*, Oxford University Press, London, 1950.
23. HOLLOMON, J. H. and TURNBULL, D. *Progress in Metal Physics*, **4** (Ed. CHALMERS, B.), Pergamon, London, 1953, p. 356: BEVER, M. B. *Energetics in Metallurgical Phenomena*, **1** (Ed. MUELLER, W. M.), Gordon and Breach, New York, 1965, p. 111.
24. DAVIES, V. de L. and WEST, J. M. *J. Inst. Metals*, **92** (1963–4), 175.
25. MORROGH, H. *J.I.S.I.* **176** (1954), 378.
26. WINEGARD, W. E. *An Introduction to the Solidification of Metals*, The Institute of Metals, London, 1964: CHALMERS, B. *Principles of Solidification*, Wiley, New York, 1964: CHADWICK, G. A. *Progress in Materials Science* (Ed. CHALMERS, B.), **12**, p. 99 (No. 2), Pergamon, London, 1963.
27. BECKER, R. *Ann. Phys.* **32** (5) (1938), 128.
28. OWEN, W. S. Theory of Heat Treatment, p. 4, of *Heat Treatment of Metals*, Iliffe, London for the Institute of Metallurgists, 1963.
29. KELLY, A. and NICHOLSON, R. B. *Progress in Materials Science* (Ed. CHALMERS, B), **10**, 151 (No. 3), *Precipitation Hardening*, Pergamon, London, 1963.
30. MEDDINGS, B. and MACKIW, V. N. Contribution to *Unit Processes in Hydrometallurgy* (Eds. WADSWORTH, M. E. and DAVIS, F. T.), *Met. Soc. A.I.M.E.*, New York, 1964, p. 345.
31. MACKIW, V. N., BENZ, T. W. and EVANS, D. J. I. *Practice and Potential of Pressure Hydrometallurgy*. I.U.P.A.C. Congress, Montreal, 1961.
32. WARD, R. G. and SALMON, K. A. *J.I.S.I.* **196** (1960), 393.
33. *Basic Open Hearth Steelmaking*, *A.I.M.E.*, New York, 3rd edn., 1964, Chapter 15 (DARKEN, L. S.), p. 593.
34. COULSON, J. M. and RICHARDSON, J. F. *Chemical Engineering*, Vol. I, Pergamon, London, 2nd edn., 1964, Chapter 3, pp. 41 and 56.
35. SZEKELY, J. *J.I.S.I.* **202** (1964), 505.
36. BRADSHAW, A. V. and RICHARDSON, F. D. Contribution to *Vacuum Degassing of Steel*, Iron and Steel Institute Special Report No. 92, 1965, p. 29.
37. DARKEN, L. S. and GURRY, R. W. *The Physical Chemistry of Metals*, McGraw-Hill, New York, 1953, p. 485.
38. MACHLIN, E. S. *Trans. A.I.M.E.* **218** (1960), 314.
39. WHITMAN, W. G. *Chem. and Met. Eng.* **29** (1923), 147.
40. RICHARDSON, F. D. *J. Inst. Metals*, **93** (1965), 525.
41. Ref. 16, p. 278.
42. SEHGAL, V. D. and MITCHELL, A. *J.I.S.I.* **202** (1964), 216.
43. WARD, R. G. *J.I.S.I.* **201** (1963), 11.
44. KOZAKÉVITCH, P. and URBAIN, G. *Rev. Mét.* **60** (1963), 143.

CHAPTER 7

Extraction and Refining of Metals

7.1. Introduction

Metals occur as *minerals* in the earth – a mineral being a naturally occurring element or compound – and minerals are the constituents of *rocks*. A rock which contains a mineral from which the metal may be extracted at a profit is called an *ore*, and the extraction metallurgist is concerned with the theory, development and control of the methods of extracting metals from ores and refining the crude extract. This chapter aims to introduce the chemistry of some of the methods in current use and to indicate how the principles outlined in earlier chapters may be applied to the solution of the "riddle of the rocks" – as it is so aptly termed in the commentary to a well-known film on nickel extraction.[1] No attempt will be made to give a descriptive treatment of the subject, and several references are available with details of plant and operation should the reader find them necessary.[2]

It has been estimated[3] that the earth's crust contains nearly 50% oxygen and over 25% silicon by weight. Aluminium, iron, calcium, sodium, potassium and magnesium follow in decreasing proportions as the next most abundant elements. It is therefore not surprising to find that whatever the metal-bearing minerals in an ore – called the *value minerals* – the oxides or compounds between the oxides of the above group of elements are likely to be present in varying quantities as impurities known as *gangue minerals.* The geological processes by which value minerals were concentrated naturally in ore deposits should be studied elsewhere[4] but usually the metal-bearing minerals are sulphides, oxides or silicates, or, where the original minerals have become chemically altered by the action of oxygen, water and carbon dioxide from the atmosphere, sulphates, carbonates and hydrated oxides. Some metals, like gold and platinum, occur "native" in important ore deposits—that is as elements rather than compounds—but this is the extreme end of a long scale of metal compounds of decreasing stability.

Many ores are not mined in a suitable form for the extraction processes, and they often do not contain a very high proportion of value minerals. Mineral dressing of some sort is always necessary, and although the concentration of value minerals by froth flotation was introduced in Section 6.4, the wide subject of mineral dressing is too important for reference to two standard texts[(5)] to be missed. If mineral dressing can be defined as processing short of chemical alteration of minerals, this chapter will take up the extraction where mineral dressing ends and chemical reactions involving bulk minerals are deliberately employed.

No mineral concentration process ever completely liberates one mineral from another, and often minerals are compounds of several metals. The problems in extraction are often associated with the behaviour of these impurities, and refining is necessary to remove their reduction products remaining dissolved in the metal after the extraction process is complete. Nevertheless, the central problem in extraction is the stability of some compound of a metal, MX. A positive free energy change ΔG will be associated with the reaction,

$$MX \rightarrow M + X,$$

and the greater the attraction of the metal for X, the greater the stability of MX, the greater the positive value of ΔG if it is to be split up, and consequently the more difficult the extraction process. Two major methods are available for separating X from M—the use of a reducing agent R whose attraction for

X is greater than that of M so that ΔG for

$$MX + R \rightarrow M + RX$$

is negative, or the solution of MX in a suitable electrolyte—aqueous or fused salt—followed by the conversion of electrical into chemical energy by electrolysis to cause the deposition of the metal at a cathode. Most other extraction reactions are subordinate to these two central reactions, but they are often aimed at improving the rate of these reactions as well as their thermodynamic possibility. Even if a reaction is accompanied by a large negative free energy change, the process will be uneconomic if the rate is too slow.

7.2. Slags

Slags, which consist primarily of oxides, are used in most *pyrometallurgical* processes—that is processes involving elevated temperatures and broadly covering extraction of metals such as iron, zinc and lead using a reducing agent such as carbon, refining by preferential oxidation (fire refining of copper and steelmaking) and matte smelting and conversion in the extraction of copper and nickel from sulphides. Slags fulfil two main functions—in extraction processes they take up the gangue minerals which are not reduced to the metallic state and in refining processes they act as the receiver for unwanted constituents of the metal. Iron provides a good example of both these functions—the slag in the iron blast furnace contains the gangue minerals such as silica, alumina and calcium oxide and, because it is liquid and separates easily from liquid iron, provides a method of removing these waste materials from the furnace separately from the liquid metal. The iron produced by the blast furnace contains up to 10% impurity elements by weight, and to convert this relatively useless material into a purer alloy, steel, oxygen is introduced into the iron. Silicon and manganese are converted into oxides which enter the steelmaking slag, and by suitable adjustment of the slag composition, sulphur and phosphorus may also be removed from the liquid metal into the slag. Extraction slags can play a refining role—as in the iron blast furnace where some sulphur is removed from the iron by the slag. Slags may also control the supply of oxygen, nitrogen, hydrogen and sulphur from the gaseous atmosphere of a reverberatory furnace to the liquid metal and can act as thermal barriers where heat is either leaving a liquid metal bath or entering it from a flame playing on the slag surface (open hearth steelmaking).

To allow these functions to be adequately fulfilled, slags must possess the following properties: they must be sufficiently fluid to allow easy separation from the metal and to increase the rate of mass transfer to and from the slag/metal interface. They must become fluid at a low enough temperature for the process to be worked economically with as little heat input and refractory wear as possible—*fluxes* such as lime, quartz, fluorspar or iron oxide may be added solely to lower the liquidus temperature and viscosity of slags. Their specific gravity must be sufficiently different from that of the metal to allow easy separation. They must have the correct composition and structure to dissolve impurities and gangue minerals at low activity and to allow any desired slag/metal reactions to occur.

In the use of slags, the following factors must be remembered: slag formation from solid constituents is usually endothermic so that the greater the mass of slag used in a process, the greater the thermal requirements of that process; the greater the volume of slag used, the greater the possibility of metal being lost by becoming mechanically trapped in the slag; rate-controlling factors in liquid slag/liquid metal reactions will include mass transfer in the slag and metal phases (Section 6.6).

The structures of molten oxides were mentioned briefly in Section 5.2, and as silica forms the basis of most slags—being the commonest gangue constituent—we can form an initial picture of slags based on two types of oxide, RO and SiO_2, where RO can represent any of the oxides CaO, MnO, MgO, FeO, ZnO, PbO, Cu_2O, Na_2O, K_2O. Ward[6] outlines a method of defining the composition of a slag in terms of the relative amounts of the two types of oxide, RO and SiO_2. RO is called a *basic oxide* because it provides oxygen ions when dissolved in a slag,

$$RO = R^{2+} + O^{2-}.$$

Silica is an *acidic oxide* which will absorb oxygen ions provided by a basic oxide,

$$SiO_2 + 2O^{2-} = SiO_4^{4-}.$$

An *acid slag* is one which contains more acidic oxide than the orthosilicate composition 2RO, SiO_2 at which each silicon atom exists as a separate SiO_4^{4-} anion. A *basic slag* contains more basic oxide than the orthosilicate composition and must therefore contain excess oxygen ions which are not part of the silicate anion structure.

In slags, other acidic oxides than silica are present, and their oxygen requirements must also be satisfied by the addition of basic oxides before the slag becomes basic. For example, the neutralization of alumina and phosphorus pentoxide can be represented by the equations

$$Al_2O_3 + 3O^{2-} = 2AlO_3^{3-}$$

$$P_2O_5 + 3O^{2-} = 2PO_4^{3-},$$

and each molecule of alumina and phosphorus pentoxide would require three molecules of basic oxides to neutralize them. This approach is useful in certain cases but it must be remembered that the picture is complicated by the tendency of some oxides (Al_2O_3, Fe_2O_3, SnO_2, ZnO and PbO) to behave *amphoterically*, that is as acidic oxides in basic slags and as basic oxides in acid slags. The extent to which aluminium is present as polymer anions in acid slags is not known, and this presents a considerable barrier to the understanding of the behaviour of iron blast furnace slags which tend to have significant alumina contents. For the sake of comparison, the *basicity* of certain extraction and refining slags has been calculated and presented in Table 7.1. The basicity is calculated as the ratio

$$\frac{(\text{Moles RO}) - 3(\text{Moles Al}_2\text{O}_3 + \text{Moles P}_2\text{O}_5)}{2(\text{Moles SiO}_2)}$$

and although the assumption that amphoteric oxides do not exist is invalid in the statement that a basicity of 1 indicates a neutral slag, it is felt that the comparison is still of some value.

In order that the activity of slag components can be calculated from slag compositions—for the purpose of predicting

TABLE 7.1. BASICITY OF SLAGS USED IN EXTRACTION AND REFINING

Process	Major slag constituents (by weight)*	Basicity†
Copper Matte Smelting	50%FeO, 38%SiO_2	0·48
Iron Blast Furnace	44%CaO, 34%SiO_2, 12%Al_2O_3, 10%MgO	0·60
Copper Converter	65%FeO, 25%SiO_2	0·98
Zinc Blast Furnace	34%FeO, 31%CaO, 16%SiO_2, 7%Al_2O_3	1·41
Basic Open Hearth Steel-making	55%CaO, 13%SiO_2, 8%P_2O_5, 7%FeO, 6%MgO	2·37

*Slag analyses are typical but considerable variation is to be expected in all processes.

†Basicity $= \dfrac{n_{RO} - 3(n_{Al_2O_3} + n_{P_2O_5})}{2n_{SiO_2}}$.

equilibria in slag/metal systems – some model of slag behaviour must be proposed. Ward[6] gives an excellent survey of slag models which have been used, and reference to Chapter 5 in his book should be made if more than the brief outline given here is required. The models are for ferrous extraction and refining because it is in steelmaking that the control of slag/metal reactions is most important and has received most attention, but there seems no reason why the treatment should not be extended to other metals than iron. The earliest models were based on the assumption that undissociated molecules formed the structure of liquid slags; for example in basic slags, the silica existed entirely as orthosilicate molecules $2RO, SiO_2$ and the remaining basic oxides were present as "free" oxide molecules CaO, MnO, etc. This approach allowed a reasonable interpretation of the behaviour of phosphorus in basic steelmaking processes. The "molecular theories" were extended by Schenck[7] to include the possibility that these molecules might be partially dissociated, for example

$$2CaO, SiO_2 \rightleftharpoons 2CaO + SiO_2.$$

"Dissociation constants" were calculated, being the apparent equilibrium constants for the dissociation reactions under the conditions existing in the slag, and hence the proportion of "free" CaO available for desulphurization of liquid iron in the blast furnace by the reaction

$$(CaO) + [S] = (CaS) + [O]$$

could be calculated for acid slags. The molecular theories eventually became so complex when they tried to account for every eventuality that it became imperative that the ionic nature of slags be recognized in any slag models. The attraction of ionic slag models, in addition to the knowledge that slags are of ionic structure, is the fact that the number of ionic species may be limited and therefore the treatment may be intrinsically less complicated. This is certainly true of basic slags, but the range of possible polymer anions in silica-rich slags has so far prevented the establishment of a self-consistent theory for acid slags.

Temkin[8] considered that a basic steelmaking slag consisted of SiO_4^{4-}, PO_4^{3-}, AlO_3^{3-} and O^{2-} anions, and the cations R^{2+} from the basic oxides, and that the slags were ideal ionic solutions. The anion and cation functions were separated assuming that the anions and cations existed separately on their own lattices. Then the activity of RO was given by

$$a_{RO} = x_{R^{2+}} \,.\, x_{O^{2-}}, \tag{7.1}$$

where the ionic fraction $x_{R^{2+}}$ is the number of cations R^{2+} divided by the total number of cations in the slag. Similarly,

$$x_{O^{2-}} = \frac{n_{O^{2-}}}{\Sigma n_{\text{anions}}}. \tag{7.2}$$

If a slag/metal equilibrium is considered, for example

$$[S] + (O^{2-}) = (S^{2-}) + [O],$$

the equilibrium constant using the ionic fractions for the slag reactants is truly a constant,

$$K = \frac{a_{[O]} \cdot x_{S^{2-}}}{a_{[S]} \,.\, x_{O^{2-}}}, \tag{7.3}$$

where $a_{[O]}$ and $a_{[S]}$ are the activities of oxygen and sulphur respectively in the metal. The Temkin model is satisfactory so long as it can be assumed that all the cations are equally important, but this is not the case – for Ca^{2+} is more important in phosphorus removal than Mg^{2+}. The theory of Flood, Førland and Grjotheim[9] uses Temkin's model but takes this difference into account, shows the importance of lime in phosphorus removal in steelmaking and is the most successful attempt to provide a model of liquid slags based on their ionic structure. They consider cation equilibrium, which is safer than anions because on the whole cations do not change their extent of polymerization or complex-formation as much as anions.

7.3. The Reduction of Oxides

One method of extracting a metal is to reduce a metal oxide MO with a reducing agent R by the reaction

$$MO + R = M + RO. \qquad (7.4)$$

For this reaction to proceed, RO must be more stable than MO, giving a negative free energy change accompanying the reaction. Preferably, the reaction should have a large equilibrium constant (Section 2.10), and by eqn. (2.39), the equilibrium constant of the reaction can be related to the standard free energy change of the reaction,

$$\Delta G^{o} = -RT \ln K_c.$$

A large equilibrium constant will give a large *negative* value of ΔG^{o}, so that a large negative value of ΔG^{o} for (7.4) will indicate a high proportion of M and RO if the reaction reaches equilibrium. Equation (7.4) can be represented as the sum of the two equations

$$MO = \tfrac{1}{2}O_2 + M \qquad (7.5)$$

$$R + \tfrac{1}{2}O_2 = RO \qquad (7.6)$$

and we know by the First Law of Thermodynamics that ΔG^{o} for (7.4) will be given by the sum of ΔG^{o} for (7.5) and (7.6) or the difference between ΔG^{o} for (7.6) and (7.7)

$$M + \tfrac{1}{2}O_2 = MO. \qquad (7.7)$$

Ellingham[10] suggested a convenient graphical method of

representing the relative stabilities of oxides. He plotted the values of the standard free energies of formation of various oxides against temperature—each standard free energy of formation involving the reaction of 1 mole of oxygen so that the distance between two curves at any temperature represented the standard free energy change accompanying the reduction of the oxide represented by the upper of the two curves by the element or compound whose oxidation was represented by the lower curve, at the temperature chosen.

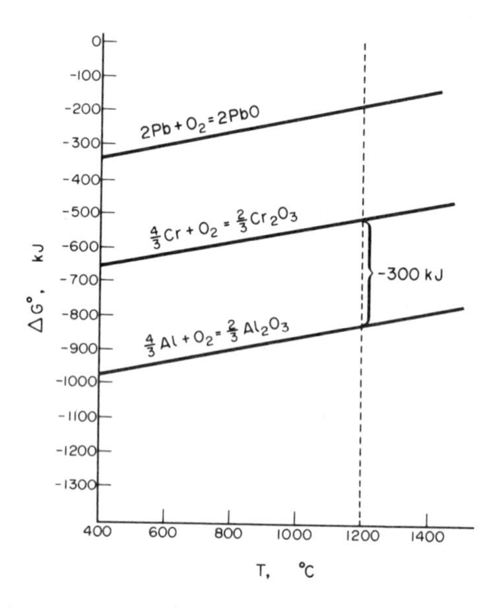

FIG. 7.1. ΔG^{o} plotted against temperature T for the reactions to form PbO, Cr_2O_3 and Al_2O_3—each from 1 mole of oxygen gas.

This can be demonstrated by reference to Fig. 7.1 at 1200°C. ΔG^{o} for the formation of $\frac{2}{3}Cr_2O_3$ is approximately −500 kJ and for the formation of $\frac{2}{3}Al_2O_3$ is approximately −800 kJ. The distance between the two lines is −300 kJ, which is the difference between these two standard free energies of formation and therefore for the reduction of Cr_2O_3 by Al.

$$\tfrac{2}{3}Cr_2O_3 = \tfrac{4}{3}Cr + O_2, \quad \Delta G^o = +500 \text{ kJ}$$

$$\tfrac{4}{3}Al + O_2 = \tfrac{2}{3}Al_2O_3, \quad \Delta G^o = -800 \text{ kJ}$$

ADD $$\tfrac{2}{3}Cr_2O_3 + \tfrac{4}{3}Al = \tfrac{4}{3}Cr + \tfrac{2}{3}Al_2O_3, \quad \Delta G^o = -800+500 = -300 \text{ kJ}$$

Thus the reduction of Cr_2O_3 by Al has a large negative standard free energy change, and because, from (2.39),

$$\ln K_c = \frac{\Delta G^o}{-RT},$$

$K_c = 4{\cdot}4 \times 10^{10}$, so that at equilibrium there will be a large proportion of chromium metal. This is the basis of the aluminothermic reduction of chrome ore, and the reaction proceeds rapidly once ignited because it is strongly exothermic and produces its own high reaction temperature. By the same argument, lead could not be used to reduce chromium oxide by the reaction

$$\tfrac{2}{3}Cr_2O_3 + 2Pb = 2PbO + \tfrac{4}{3}Cr$$

because ΔG^o is approximately +330 kJ and K_c is correspondingly very small indeed. We can therefore see the use of the *Ellingham diagram* to indicate which reducing agents might possibly be used to reduce an oxide, and to give a graphical impression of the relative stabilities of a number of oxides. Ellingham diagrams have been produced for sulphides, sulphates, carbonates, chlorides and fluorides in addition to those for oxides, and as more thermochemical data have become available, they have been extended and brought up to date by various workers.[(11)]

Figure 7.2 is an Ellingham diagram for oxides (after that by Richardson and Jeffes[(11)]) and will serve to illustrate some of the points to be made in this chapter on the relative stability and ease of reduction of metal oxides. The shape and slope of the curves is worthy of some consideration, as this is determined by the entropy change accompanying each reaction.

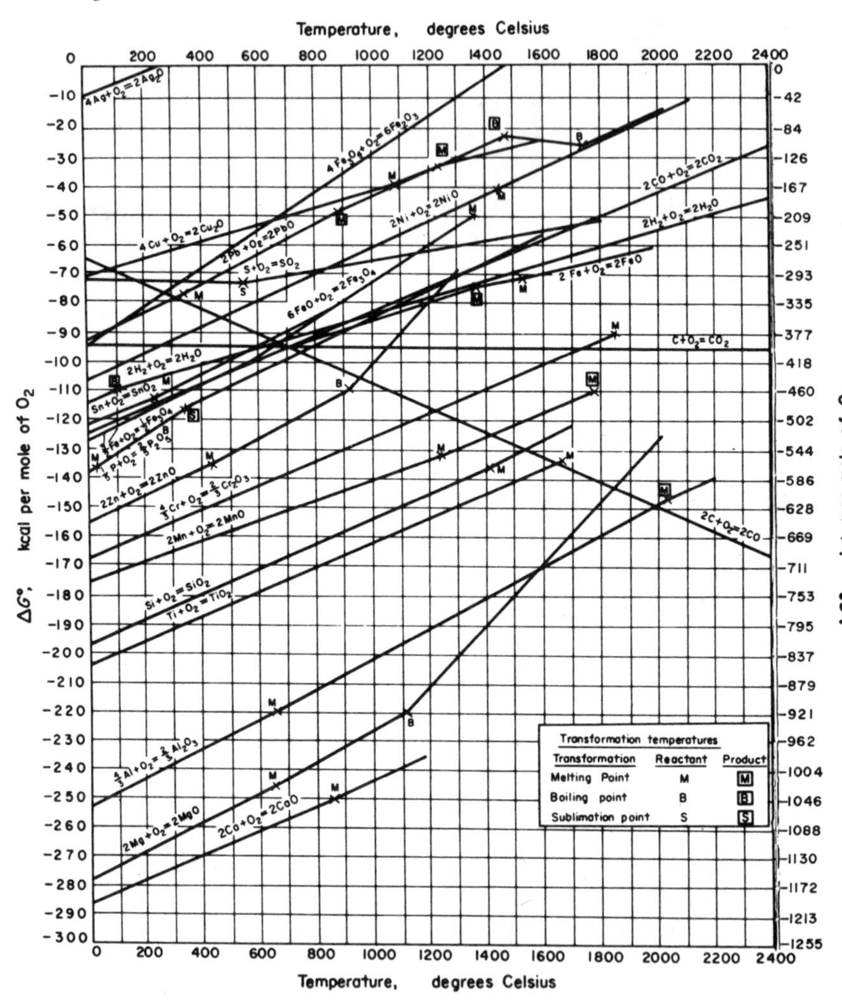

FIG. 7.2. The standard free energy of formation of a number of oxides as a function of temperature (Ellingham diagram). Accuracy varies between ±4 and ±40 kJ. [After Richardson, F. D. and Jeffes, J. H. E. *J.I.S.I.* **160** (1948), 261.]

From (2.18)

$$\Delta G^{\circ} = \Delta H^{\circ} - T\Delta S^{\circ},$$

so that as ΔG° is plotted as a function of T in these diagrams, $(-\Delta S^{\circ})$, the standard entropy change of reaction, should be the slope of the curve. Although ΔS° and ΔH° are temperature-dependent, so long as the reactants and products stay in the same physical state the variation of ΔS° and ΔH° with T is small and the curves are virtually straight lines. Most of the reactions are of the form

$$M_S + O_{2_G} = MO_S$$

at low temperatures, so that most of the ΔS° values are approximately the same (negative because of the increase in the state of order of the system due to the loss of 1 mole of oxygen gas) and most of the lines have roughly the same positive slope. When the metal changes phase (e.g. at the melting point or boiling point shown as M or B on Fig. 7.2), ΔS° will be smaller so that $-\Delta S^{\circ}$, the slope of the line, will become more positive. If the oxide changes phase ($\boxed{\mathrm{M}}$ or $\boxed{\mathrm{B}}$ on Fig. 7.2), ΔS° will be larger and therefore the slope more negative. This is shown clearly by the line for zinc

$$2Zn + O_2 = 2ZnO,$$

where there is a slight increase in the slope of the line at the melting point and a large increase at the boiling point of zinc. This illustrates the discussion in Section 2.2 in which it was pointed out that the entropy change (10·5 J/°K) accompanying the melting of zinc was much smaller than the entropy of evaporation (96 J/°K). The important exceptions to the general tendency to have positive slopes on Fig. 7.2 are the two lines for the oxidation of carbon,

$$C_S + O_{2_G} = CO_{2_G} \tag{7.8}$$

and $$2C_S + O_{2_G} = 2CO_G. \qquad (7.9)$$

ΔS° for (7.8) is about $+0{\cdot}8$ J/°K because one gas molecule appears on either side of the equation and there is consequently only a small change in entropy accompanying the reaction. The line for (7.8) on Fig. 7.2 is consequently almost horizontal. ΔS° for (7.9) is about $+170$ J/°K because there is an increase in the number of gas molecules accompanying the reaction. The line for (7.9) has therefore a pronounced negative slope ($-\Delta S^\circ$), and this is of great significance in metal extraction because it means that if the temperature is raised, CO becomes more and more stable with respect to metal oxides, and therefore carbon will reduce more and more metal oxides as the temperature is raised. ΔG° for

$$PbO + C = Pb + CO$$

is negative above 310°C, but ΔG° for

$$2PbO + C = 2Pb + CO_2$$

is negative above 0°C, so that (7.8) is the equation to be considered below about 700°C [where the lines for (7.8) and (7.9) cross]. Above 700°C, CO becomes more stable than CO_2, and will predominate in the products of any reaction in which carbon is used as a reducing agent. ΔG° for

$$FeO + C = Fe + CO$$

becomes negative above 720°C, and for

$$ZnO + C = Zn + CO$$

above 950°C. This means that in any process using carbon as a reducing agent, the thermodynamic evidence suggests that at temperatures above 0°C, large amounts of lead will appear at equilibrium in the reduction of lead oxide, but the temperature must exceed 720°C for iron and 950°C for zinc to be produced in significant amounts. This is the reason for zinc vapour (b.p. of zinc is 907°C) appearing as the product of carbon reduction of zinc oxide in retorts or the blast furnace as was pointed out in Section 2.6. Zinc smelting using carbon must be carried out at temperatures of the order of 1100°C if a significant rate of reduction of ZnO is to be achieved.[12]

This argument can be extended to the possibilities of reducing stable oxides such as TiO_2, MgO and Al_2O_3, but ΔG^o becomes negative for the reduction of these oxides by carbon above 1630°C, 1840°C and 2000°C respectively according to Fig. 7.2, and to operate smelting processes above these high temperatures would be expensive because of the fuel requirements and the difficulty of producing refractory containers which would last for an economic period. In addition, the metals produced would be highly reactive – picking up oxygen from oxide refractories and from the oxides of carbon and reacting with carbon itself in the case of titanium and aluminium. As a consequence of these difficulties, titanium is extracted by reduction of $TiCl_4$ by magnesium, aluminium by electrolysis of Al_2O_3 dissolved in fused fluorides, and magnesium by electrolysis of the fused chloride, rather than by conventional carbon reduction.

Ellingham diagrams should be used with some care, because they can only show a definite trend where two lines are a reasonable distance apart. The accuracy of the data from which they are plotted and the difficulty in reading the distance between two closely spaced curves do not permit a definite conclusion where lines are much less than about 20–40 kJ apart. In addition to this there are two important drawbacks:

1. ΔG^o is the standard free energy change of the reaction and does not take into account activities of reactants or products which are significantly different from unity.

2. No account of the kinetics of the reaction is taken by measurements of thermodynamic variables.

The problem of deviations from unit activity is illustrated by the reduction of MgO with silicon. According to Fig. 7.2, ΔG^o for

$$2MgO_S + Si_S = 2Mg_G + SiO_{2_S}$$

at 1200°C is about +272 kJ, so that there appears to be very little chance of using silicon as a reducing agent to produce magnesium from magnesia. However, consideration of the van't Hoff isotherm (2.37) and (2.39) and the discussion of activities in Section 3.4 shows that the actual free energy change accompanying this reaction is given by

$$\Delta G = \Delta G^o + RT \ln \frac{p_{Mg}^2 \cdot a_{SiO_2}}{a_{MgO}^2 \cdot a_{Si}}.$$

If p_{Mg} and a_{SiO_2} can be lowered sufficiently, ΔG can be made negative even though ΔG^o is positive. The Pidgeon process[(13)] for the commercial production of magnesium lowers p_{Mg} by working at a pressure of about 10^{-4} atm and by having CaO present to lower the activity of the silica in a slag as the orthosilicate 2CaO, SiO_2. (A basic slag would give $a_{SiO_2} < 0{\cdot}001$.) The strong attraction of CaO for SiO_2 reduces the possibility of loss of MgO as magnesium silicate. A convenient method of introducing CaO is to use calcined dolomite, which is a mixed carbonate of magnesium and calcium, as the source of magnesium,

$$\underset{\text{dolomite}}{CaCO_3, MgCO_3} \xrightarrow{\text{heat}} CaO, MgO.$$

$$2(CaO, MgO) + Si = 2Mg_G + 2CaO, SiO_2.$$

In the Pidgeon process, ferrosilicon is used rather than pure silicon and the iron is dissolved as FeO in the slag—lowering the liquidus temperature of the slag. The magnesium is evolved as a gas, which is condensed in massive form without reoxidation. This should be compared with the difficulties in the reduction of MgO with carbon in which the carbon monoxide evolved with the magnesium vapour reoxidises the magnesium on cooling. This means that the magnesium vapour must be shock-cooled with cold hydrogen, forming a finely divided, pyrophoric magnesium powder which is difficult to handle,

$$MgO_S + C_S \overset{1900°C}{=} Mg_G + CO_G.$$

From this example alone, it should be clear that if the activities of reactants or products differ significantly from unity, the van't Hoff isotherm (2.37) gives a better indication of the thermodynamic possibilities of the reaction than ΔG^o and the Ellingham diagram. The condensation of metallic vapours in the presence of the oxides of carbon is always a problem in extraction metallurgy, particularly in the extraction of zinc by carbon reduction of ZnO, where the reaction

$$Zn_G + CO_G = ZnO_S + C_S$$

and, more significantly,

$$Zn_G + CO_{2_G} = ZnO_S + CO_G$$

will occur when the vapour is cooled below 900°C. This problem has been solved very neatly in the Imperial Smelting Process[14] where zinc is produced by carbon reduction of ZnO in a blast furnace. The product gas, containing zinc, carbon monoxide, carbon dioxide and nitrogen is kept at a temperature of about 1000°C by a controlled addition of air, which reacts with CO exothermically

$$CO + \tfrac{1}{2}O_2 = CO_2, \qquad \Delta H = -280 \text{ kJ}.$$

The gas is then passed into a "lead splash condenser", where liquid lead at 560°C is agitated violently and is used to condense and dissolve the zinc vapour without significant re-oxidation. The liquid lead (containing about 2·4% zinc in solution) is removed from the condenser and cooled to 450°C, at which temperature the maximum solubility of zinc is 2·15%. The superfluous zinc (98·5% pure) comes out of solution and floats on the surface of the lead, whence it can be removed; this process is called *liquation.* The lead is then introduced in closed circuit into the splash condenser (Fig. 7.3). The development of the Imperial Smelting Process represents an excellent example of the successful application of physico-chemical principles to a commercial process.

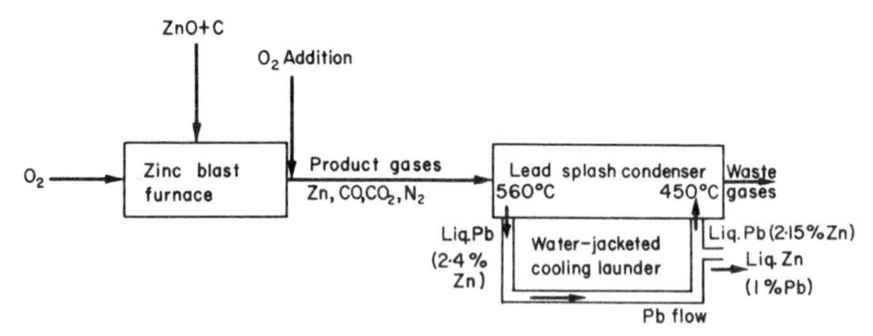

FIG. 7.3. Flow sheet showing the condenser unit in the Imperial Smelting Process.

Reaction kinetics cannot be ignored in any process (Chapters 4 and 6), and the example of the lead blast furnace process is worth considering in more detail. From a thermodynamic aspect,

$$2PbO + C = 2Pb + CO_2$$

will take place at 100°C, because ΔG is negative. A blast furnace (Fig. 7.4) charged with solid lead oxide, coke and lime flux would remain virtually unaltered at 100°C because the rate of chemical reactions would be too slow. If air (preheated by burning the CO in the product gases in some form of heat

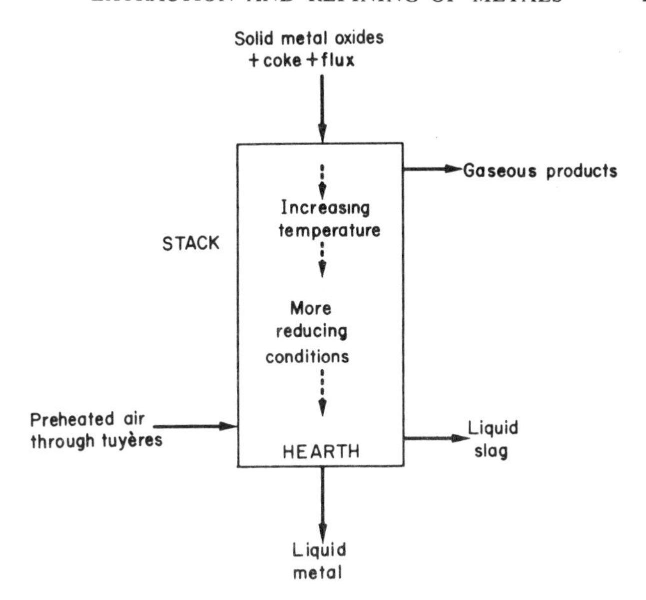

FIG. 7.4. Representation of a blast furnace process.

exchanger) is introduced via the tuyères (a form of blow pipe), the coke will burn in the tuyère zone with the release of heat

$$C + O_2 = CO_2, \qquad \Delta H = -393 \text{ kJ}. \qquad (7.8)$$

The CO_2 will react with more coke to produce carbon monoxide

$$CO_2 + C = 2CO, \qquad \Delta H = +167 \text{ kJ}, \qquad (7.10)$$

to give an overall reaction, the sum of (7.8) and (7.10), using Hess's law (Section 1.9),

$$2C + O_2 = 2CO, \qquad \Delta H = -226 \text{ kJ}. \qquad (7.9)$$

Equation (7.9) is the most likely overall reaction at the tuyères because the equilibrium constant of (7.10) is very high at high temperatures – giving a predominance of CO over CO_2 with C present in excess. The exothermic nature of the reactions (7.8) and (7.9) is one reason for the widespread use of carbon as a reducing agent – it supplies enough heat when burnt in oxygen to keep the blast furnace process going without the necessity for additional sources of heat. This heat is radiated to the materials in the hearth (slag and metal) and is transmitted via the rising hot gases to the charge in the stack – resulting in an increase in temperature in the charge as it moves down the stack. The temperature in the lead blast furnace is thus raised to about 1200°C at the top of the hearth, giving a rapidly formed liquid lead product which separates easily from the liquid slag produced by the gangue minerals and flux, and can be tapped separately from the furnace. Because lead oxide is reduced high in the furnace and melts well up in the stack (top gas temperature is 200°C), the volume of the charge decreases steadily from top to bottom, and an inward-sloping furnace wall has been used to give a smooth flow of material.

Although thermodynamically the reduction reaction in a process using carbon as a reducing agent is represented as

$$MO_S + C_S = M + CO_G \qquad (7.11)$$

or

$$2MO_S + C_S = 2M + CO_{2_G}, \qquad (7.12)$$

where M may be solid, liquid or gaseous, depending on the temperature of the reaction, the *rate* of a reaction between two solids, MO and C, is likely to be very slow as it will be controlled by the rate of diffusion in the solid state across a small area of contact. It is more probable that the reaction takes place in more than one step, involving gas/solid reactions which are likely to be more rapid, for example

$$MO_S + CO_G = M + CO_{2_G}, \qquad (7.13)$$

$$CO_{2_G} + C_S = 2CO_G. \qquad (7.10)$$

Equations (7.13) and (7.10) added together produce the overall reaction (7.11). Below 700°C, the reaction will be more probably

$$MO_S + CO_G = M + CO_{2_G} \qquad (7.13)$$

alone, the result of reduction by carbon monoxide (produced at the tuyères) passing rapidly up the furnace stack. The possibility of reduction by CO alone can be examined thermodynamically by considering the relative position of the lines in Fig. 7.2

$$2M + O_{2_G} = 2MO_S \qquad (7.7)$$

and

$$2CO_G + O_{2_G} = 2CO_{2_G}. \qquad (7.14)$$

It will be seen that (7.14) lies below the lines

$$2Pb + O_2 = 2PbO$$

and

$$2Fe + O_2 = 2FeO$$

at all temperatures below 700°C, and therefore (7.13) is to be expected in the iron and lead blast furnaces, at those temperatures. The temperature should not be allowed to fall to 700°C in the zinc blast furnace so that in this case the question does not arise. Reactions of the type (7.11) are known as "direct" reduction and (7.13) "indirect" reduction and a full discussion of this subject would be beyond the scope of this book. The reader can consult other references[15, 16] for a more complete treatment of a controversial subject. Certainly, as the gases pass up the stack, the ratio p_{CO}/p_{CO_2} will decrease and the gases will become less reducing. In the case of the iron blast furnace, the oxides Fe_2O_3, Fe_3O_4 and FeO will be present and will be reduced in that order at increasing p_{CO}/p_{CO_2} ratios (see Section 2.10). As the gases cool, the chance of reaction (7.10) being reversed will be increased because the system will have excess

CO present. We saw in Section 4.11 that the reaction may be catalysed by iron compounds, so that carbon deposition may be expected below about 900°C, with a consequent increase in the CO_2 content of the furnace gases in the range 450–750°C.

Reaction (7.8) is accompanied by a larger release of heat energy than reaction (7.9), so that the ratio of CO/CO_2 in the product gases at the top of the furnace gives an indication of the *thermal efficiency* of the process. If the ratio is high, carbon will have been burnt inefficiently and an increase in coke consumption is to be expected. In the zinc blast furnace, the ratio is kept as low as possible, without causing reoxidation of zinc vapour, to give maximum thermal efficiency. Ratios must be higher at equivalent temperatures in the iron blast furnace because a higher CO/CO_2 ratio is required to reduce FeO than ZnO above about 950°C and the reduction of FeO in the zinc blast furnace should be avoided in any case. PbO can be charged in the zinc blast furnace because it will easily be reduced to lead under these conditions, and liquid lead can be tapped separately from the base of the furnace.

The structure of metallurgical coke is important in blast furnaces as it is strong enough to hold the weight of the burden without shattering, yet porous and able to promote good gas transport in the stack. Solid coal would present too small a surface area for efficient heat and mass transfer between carbon and gases. Lower in the furnace, when liquid slags have begun to form, there is probably some direct reaction of the solid coke with liquid oxides dissolved in the slag, for example

$$(ZnO)_L + C_S = Zn_G + CO_G,$$

$$(SiO_2)_L + 2C_S = [Si]_L + 2CO_G.$$

Reactions of the latter type can result in silicon dissolving in the liquid iron as an impurity in the iron blast furnace.

The blast furnace has been covered in greater detail because of its economic importance and because its operation involves some important chemical principles, but there are many other processes in which carbon is used as a reducing agent. Examples of these include zinc extraction in horizontal and vertical retorts, the production of tin in reverberatory furnaces

$$SnO_2 + 2C = Sn + 2CO,$$

and ferroalloys in electric arc furnaces,

$$\underset{\text{Chromite ore}}{FeO, Cr_2O_3} + 4C = \underset{\text{Ferrochrome}}{Fe, 2Cr} + 4CO.$$

Finally, hydrogen has been used as a reducing agent in the extraction of tungsten at about 800°C,

$$WO_{3_S} + 3H_{2_G} = W_S + 3H_2O_G,$$

and in alternatives to the blast furnace for the production of iron. (The position of the hydrogen line of the Ellingham diagram suggests that the equilibrium constant for the reaction

$$FeO + H_2 = Fe + H_2O$$

is not very favourable, but it becomes more favourable as the temperature increases.) A convenient method is to use hydrogen and carbon monoxide, produced by the "water gas reaction" when steam is passed over red hot coke

$$C + H_2O = CO + H_2,$$

as reducing agents—for example in the Wiberg-Söderfors process[17] in which CO and H_2 are passed into a kiln containing high grade iron ore or concentrates at about 1000°C. The product is solid "sponge iron", which still contains the gangue oxides (there is no liquid slag formation to remove these impurities), and this is used as the charge for electric-arc steelmaking. Equilibrium will not be reached in processes using only hydrogen and carbon monoxide for reducing iron oxide and the production of significant amounts of iron will depend on the gaseous products of the reduction reactions being swept away from the sites of the reactions. This is a clear case in which Fig. 7.2 alone will not give an accurate picture of the process, because above 700°C, both the $H_2 \rightarrow H_2O$ and $CO \rightarrow CO_2$ lines lie below the $Fe \rightarrow FeO$ line, yet a commercial process is operating at 1000°C.

7.4. Thermal Pretreatment

We have seen in the last section that oxides are convenient starting-points for extraction. Metals often occur as sulphides, carbonates, hydrates or basic carbonates, and these are not always suitable, for example carbon and hydrogen cannot be used to reduce metal sulphides because their sulphides are less stable than the sulphides of the common metals; for

$$2ZnS + C = 2Zn + CS_2,$$

ΔG^0 is +439 kJ at 0°C and still +251 kJ at 900°C. These minerals are not soluble in water, so that if an aqueous solution of the metal is required it might be necessary to convert them into a more easily soluble form. Therefore ores and concentrates are often altered chemically by a thermal pretreatment, for example to produce oxides which are suitable for carbon reduction or sulphates which are water-soluble.

The simplest treatment is *ore drying,* which removes water which is physically combined with an ore or concentrate. Water boils at 100°C so that temperatures above this are only necessary to give a rapid drying process. *Calcining* may be defined as heating to produce a chemical decomposition, for example

$$CaCO_{3_S} = CaO_S + CO_{2_G}$$

$$Al_2O_3, 3H_2O_S = Al_2O_{3_S} + 3H_2O_G.$$

The temperature to which carbonates and hydrates (in which the water is chemically combined) must be raised to produce decomposition can be calculated by measurement of their "dissociation pressures" (Section 2.11) at various temperatures. The temperature above which p'_{CO_2}, the partial pressure of CO_2 in equilibrium with a carbonate, exceeds the partial pressure of CO_2 in the atmosphere in contact with the carbonate will be the minimum temperature for calcining that carbonate. The same applies to values of p'_{H_2O} for hydrates. Calcium carbonate dissociates between 700°C and 800°C, and this reaction is important in the blast furnace where

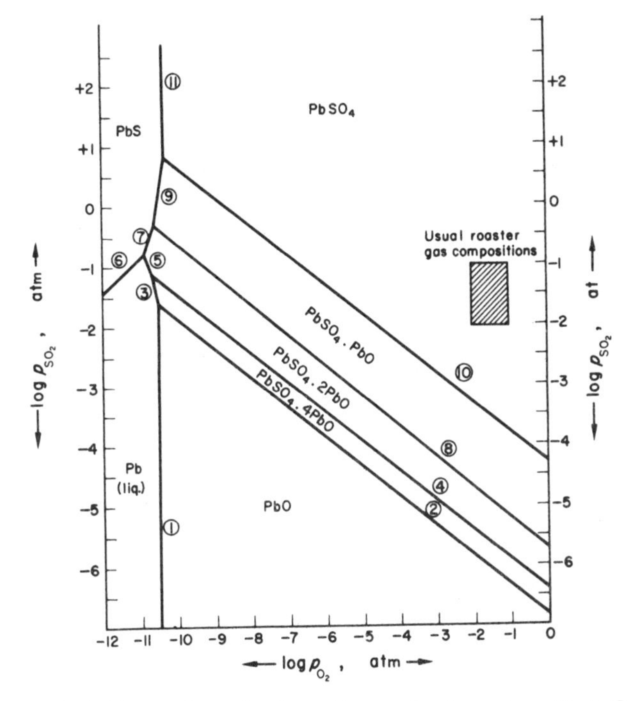

FIG. 7.5. Thermodynamic phase diagram for the system Pb–S–O, showing bivariant equilibria at 1100°K. Numbers in circles refer to the equilibria described in the text. (After Kellogg, H. H. and Basu, S. K.[18])

limestone is used as a flux because at this temperature CO_2 will be released and the CO/CO_2 ratio in the furnace atmosphere will decrease. Other metal carbonates may dissociate at lower temperatures—for example, iron and zinc carbonates calcine above 400°C. Hydrates will usually dissociate at temperatures at which the corresponding carbonates dissociate, and the temperatures used in calcining of gibbsite (Al_2O_3, $3H_2O$) during the purification of alumina in the Bayer process (prior to the fused salt electrolytic extraction of aluminium) is high (1250°C) because this ensures that the alumina remains in the α form, rather than producing $\gamma-Al_2O_3$ which adsorbs water vapour too easily and would carry water into the electrolytic cell where the water would react with the fluoride electrolyte to form HF. Rates of calcining reactions will depend on the surface area of the solid and often on heat transfer through the calcined layer in large lumps of rock because they are endothermic reactions.

Roasting involves chemical combination with the atmosphere, and is usually associated with the heating of sulphide minerals to cause a reaction with oxygen in the atmosphere. Reactions occurring may include

$$2MS+3O_2=2MO+2SO_2 \qquad (7.15)$$

$$MS+2O_2=MSO_4 \qquad (7.16)$$

$$MS+O_2=M+SO_2 \qquad (7.17)$$

and

$$MS+O_2 \rightarrow MSO_4 \cdot xMO. \qquad (7.18)$$

The process may require an external source of heat or may be autogenous, depending on the exothermic nature of the reactions to provide the necessary heat. A convenient method of demonstrating the conditions favouring the various reactions is provided for the roasting of lead sulphide by Kellogg and Basu[18] who examined equilibria for the reactions involving the various phases produced, and plotted a thermodynamic "phase diagram" for the Pb–S–O system at 827°C (Fig. 7.5). The lines on the diagram show the relationship between $\log p_{SO_2}$ and $\log p_{O_2}$ at equilibrium for the reactions, assuming that SO_2 and O_2 would be the only gaseous reactants taking part in any of the possible reactions. SO_2 will always be present as soon as oxygen has reacted with a sulphide, so that its concentration must be taken into account. Referring to Fig. 7.5, lines 1 and 11 are vertical because they represent equilibria in which SO_2 plays no part.

1. $2PbO_S = 2Pb_L + O_{2_G}$.

11. $PbS_S + 2O_{2_G} = PbSO_{4_S}$.

The equilibrium constant is p'_{O_2} in 1, $1/p'^2_{O_2}$ in 11, assuming unit activity of solid and liquid reactants. Referring to 11, if $\log p_{SO_2}$ is greater than 0·83 and $\log p_{O_2}$ lies to the right of line 11, lead sulphide will be roasted to lead sulphate. Lines 3, 5, 7 and 9 refer to the equilibria involving the formation of basic sulphates from liquid lead and lead sulphide, 7 and 9 being of greater importance in roasting.

3. $5Pb_L + 3O_{2_G} = PbSO_4, 4PbO_S$.

5. $3Pb_L + SO_{2_G} + 2O_{2_G} = PbSO_4, 2PbO_S$.

7. $3PbS_S + 5O_{2_G} = PbSO_4, 2PbO_S + 2SO_{2_G}$.

9. $2PbS_S + \frac{7}{2}O_{2_G} = PbSO_4, PbO_S + SO_{2_G}$.

The equilibrium constant of each of these reactions involves both p'_{SO_2} and p'_{O_2}, so that variation in p_{O_2} will cause a variation in p_{SO_2} at equilibrium and vice versa. If lead sulphide was introduced into a chamber containing gases in which $\log p_{SO_2}$ was −2 and $\log p_{O_2}$ was −8, the basic sulphate $PbSO_4, 2PbO$ would be the roast product. Lines 2, 4, 8 and 10 refer to equilibria between the sulphate, the basic sulphates and the oxide. Thus if p_{SO_2} is gradually decreased at constant p_{O_2}, $PbSO_4$ will react to form the basic sulphates and finally become pure PbO at very low values of p_{SO_2}. Finally, line 6 is the so-called *roast-reduction reaction*

6. $PbS_S + O_2 = Pb_L + SO_2$

in which lead sulphide can be converted to metallic lead by a careful control of p_{SO_2} and p_{O_2}. In fact at this temperature the values of p_{SO_2} and p_{O_2} necessary to produce liquid lead would be too low for the conditions usually occurring in the roaster, but at higher temperatures the Pb field expands upwards and to the right until the roast reduction reaction becomes a possibility and liquid lead appearing in the roaster can be a nuisance. Lead may be produced by this reaction at high temperatures (involving PbS left unreacted during roasting) in the blast furnace or in the old ore-hearth process.

One important implication of Kellogg and Basu's results is that at no point is there a common boundary between the PbS and PbO fields. This means that PbO cannot be produced by roasting PbS in oxygen without some sulphate formation. The roasted particle will consist of PbO in the outer layers, where p_{SO_2} is kept very low by sweeping the SO_2 away with excess air, then layers of basic lead sulphates and even possibly $PbSO_4$ alone next to the inner zone of unroasted sulphide. Diffusion will be slow through the sulphate layers and it will be difficult to achieve complete roasting, even to the sulphate. The usual roaster gas compositions are shown in Fig. 7.5, and even with excess oxygen, the difficulty of "dead roasting" to the oxide is obvious. A "sulphatizing" roast to produce $PbSO_4$ would be relatively easily achieved by restricting the amount of oxygen present, but $PbSO_4$ is insoluble in water and would be no use for leaching purposes. The effect of raising the temperatures is to raise lines 2, 4, 8 and 10 and therefore improve the chance of producing PbO. The *sintering machine*, in which air is drawn through a hot bed of sulphide concentrates, is an ideal method of sweeping the SO_2 product away and increasing the amount of PbO formed. Care must be taken not to allow the temperature to become too high, otherwise reaction 6 will proceed and liquid lead is carried off into the fan system. The product of the sinter machine is an agglomerated, porous mass, having sufficient strength and surface area to make it an ideal charge for the blast furnace.

Zinc sulphide is roasted on sinter strands to produce the oxide for carbon reduction and in other roasting units to produce the sulphate for leaching prior to extraction by aqueous electrolysis. Copper sulphide concentrates may be given a sulphatizing roast to prepare them for leaching or an oxidizing roast aimed mainly at converting the iron sulphide gangue minerals present to oxide prior to matte smelting. The principles expounded for lead sulphide roasting still hold for other sulphides—a roast to produce oxides is favoured by high temperatures (800–900°C) and a low p_{SO_2} (excess air), whereas sulphatizing is favoured by low temperatures (600–700°C) and restricted air supply with possible recirculation of product gases. All roasting reaction rates will be dependent on a large gas/solid interfacial area and therefore a finely divided solid charge with good stirring will be advantageous. Flash or suspension roasting and fluosolids roasting make use of rapid reactions when fine sulphide particles are suspended in air.[19,20]

Roasting of sulphides on a sinter machine may require no other source of fuel because the reactions are strongly exothermic, but iron oxide ores are mixed with coke breeze, which burns on the sinter strand to cause incipient fusion of the ore particles. This process was originally carried out to agglomerate iron ore fines, but modern blast furnaces are now using a high proportion of sinter feed because it has been found that iron production is increased and made more consistent as a result of certain changes taking place on the sinter strand. Better blending of ores, the partial formation of a slag (if fluxes are incorporated in the sinter mix) which only has to be remelted in the blast furnace, and the calcination of carbonates and hydrates more efficiently on the sinter strand than in the blast furnace are benefits in addition to the production of a strong, porous agglomerate.[21] Sinter is unsuitable for very fine particles and has been joined by pelletizing as an agglomeration process for fines produced, for example, after froth flotation and magnetic concentration of taconite deposits in the U.S.A. and for the utilization of flue dust. A binder is mixed with the ore or concentrate, a little moisture is added and the mixture is rolled or extruded into pellets which are baked in some sort of furnace or on a grate such as is used in sintering. Claims have been made that better strength, reducibility and size range are produced than with a comparable sintering process.[22]

7.5. Smelting of Sulphides

Liquid oxides (slags) and liquid sulphides (mattes) are immiscible, and can be made to separate into two layers if of sufficiently low viscosity and different specific gravity. This, coupled with the relative stabilities of various oxides and sulphides, is the basis of *matte smelting*, the first stage in the smelting of copper sulphide concentrates.[23] A mixture of copper sulphide, iron sulphide, siliceous gangue minerals, iron oxide, copper oxide and impurity minerals, produced by roasting of a concentrate to convert some iron sulphide to oxide, is charged into a reverberatory furnace at about 1300°C. Flux (silica or lime, depending on the nature of the gangue) and converter slag (high iron oxide content) are added and a slag is formed, containing most of the oxides present (see Table 7.1, p. 268). The sulphides form a matte, which is undoubtedly ionic in structure, and which separates from the slag as it has a greater specific gravity (5) than that of the slag (3–4). The iron–oxygen attraction is stronger than the attraction of oxygen for copper (ΔG^o for Cu_2O formation is less negative than that for FeO – see Fig. 7.2, p. 274) so that any copper dissolved in the slag or mixed as oxide in the matte will be displaced by a transfer which can be represented as

$$FeS + Cu_2O = FeO + Cu_2S, \qquad \Delta G^{\circ} = -230 \text{ kJ},$$

whose equilibrium constant will be made more favourable for copper transfer into the matte by the solution of FeO at low activity in the siliceous slag. In this way, virtually all the copper enters the matte, which is a mixture of Cu_2S and FeS, containing most of the precious metals (Ag, Pt) which can be recovered at a later stage. The amount of sulphur present is slightly less than is necessary to make up the stoichiometric amount of Cu_2S and FeS, so that some metal must be present in the free state, and would separate out on slow cooling of the matte. The iron content of the matte can be lowered by roasting the concentrates prior to matte smelting for a longer period to produce more iron oxide, but if too rich a matte is produced, which has too little iron sulphide present, the subsequent converter process will be short of heat energy released when FeS is converted to the oxide,

$$2FeS + 3O_2 = 2FeO + 2SO_2, \qquad \Delta H = -921 \text{ kJ}.$$

A copper content of about 40 wt.% is the usual aim for a suitable matte. The heat released in the last reaction is used in the flash smelting process developed by the International Nickel Company in Canada and at Outokumpu in Finland. Dried concentrate is used as the fuel for a pulverized fuel burner using tonnage oxygen for combustion, and the formation of the matte occurs as the burning iron sulphide falls with the copper sulphide particles into a bath of slag and matte. No other fuel is necessary and the output of the furnace is higher than the conventional reverberatory furnace.[24]

Matte smelting is really a concentration process in preparation for the *converter process* of copper extraction. Liquid matte is charged into a side-blown converter with silica flux and some scrap copper, if sufficient heat energy is to be released in the process to melt the scrap. Air is blown through the matte, iron is preferentially oxidized and enters the siliceous slag as an oxide (Table 7.1). Slag is removed and more matte charged periodically until virtually all the iron has gone and the converter is full of liquid copper sulphide ("white metal"). Conditions are now suitable, with the temperature raised above 1250°C by the exothermic oxidation of iron, for the "roast reduction" reaction, similar to reaction 6 on Fig. 7.5 in the last section,

$$Cu_2S + O_2 = 2Cu + SO_2.$$

(This reaction is sometimes thought to occur in two stages, namely

$$2Cu_2S + 3O_2 = 2Cu_2O + 2SO_2$$

and $$2Cu_2O + Cu_2S = 6Cu + SO_2.$$

This would produce the same overall reaction but the single-stage reaction seems much more favourable kinetically.) Reduction is continued until the charge has been converted to "blister copper"—98% Cu containing any of the following impurities: Ni, Co, Fe, Sn, Sb, As, Zn, Pb, S and the precious metals. This copper must be refined further before it is of commercial use.

Copper–nickel–iron sulphide concentrates are treated similarly to copper concentrates at International Nickel Company's Sudbury, Ontario, plant, but the converter process is stopped after the removal of iron. The liquid "Bessemer Matte", containing copper and nickel sulphides, is 10% deficient in sulphur so that there is also present in effect a copper–nickel alloy, which has almost all the platinum metals dissolved in it. The matte is allowed to cool very slowly so that a good separation of copper and nickel sulphides is achieved.[25] The solid matte is crushed and the copper–nickel alloy containing platinum metals is separated magnetically. The copper and nickel sulphides are separated by froth flotation to complete a carefully planned operation.[1]

7.6. The Advantages of Halides

Although halides of metals do not occur frequently as value minerals, the conversion of oxides into halides can provide an attractive alternative route for metal extraction. Chlorination may be sufficient, or leaching with hydrochloric acid, but often the halides of the reactive metals are very much less stable than their oxides so that carbon may be used as a reducing agent to remove the oxygen. For example, ΔG^0 for

$$TiO_2 + 2Cl_2 = TiCl_4 + O_2$$

is $+146$ kJ, whereas ΔG^o for

$$TiO_2 + 2Cl_2 + C = TiCl_4 + CO_2$$

is about -251 kJ at 500°C, at which temperature chlorination is carried out. Chlorination of MgO is carried out in the presence of carbon for the same reason.

Titanium is a very reactive metal, occurring naturally as the oxide. We saw in Section 7.3 that carbon reduction of the oxide would be difficult, if not impossible, in a commercial process, so that the properties of $TiCl_4$ are exploited in the Kroll process[26] in which magnesium is used as a reducing agent at 850°C,

$$TiCl_{4_G} + 2Mg_L = 2MgCl_{2_L} + Ti_S.$$

$TiCl_4$ is a volatile compound boiling at 140°C, and a relatively pure form can be condensed from the gaseous products of the chlorination of TiO_2 in the presence of carbon. The titanium is produced as a relatively inert solid "sponge", rather than a reactive liquid, and the magnesium chloride and excess magnesium are removed by vacuum distillation. If magnesium were used as a reducing agent for TiO_2, the rate of the liquid/solid reaction would be comparatively low and the solid MgO produced would be difficult to separate from the titanium. Sodium has also been used by Imperial Chemical Industries to reduce $TiCl_4$, and zirconium has been extracted by the decomposition of ZrI_4 vapour at 1400°C (the Van Arkel process),

$$ZrI_{4_G} = Zr_S + 2I_{2_G}.$$

Titanium and zirconium must be vacuum-arc melted into a water-cooled copper crucible to prevent contamination. Uranium is extracted by the reduction of its fluoride by calcium or magnesium.

The volatility of halides has been used in a process which has been suggested as an alternative to the electrolytic extraction of aluminium. Impure bauxite is smelted in an electric arc furnace with carbon to produce an Fe–Si–Al alloy. $AlCl_3$ gas is passed over briquettes of this alloy above 1000°C, and the sub-halide of aluminium is produced by the reaction[27]

$$AlCl_{3_G} + 2[Al] = 3AlCl_G.$$

On cooling, the sub-halide decomposes, depositing aluminium powder and leaving $AlCl_3$ vapour which may be re-used,

$$3AlCl_G = 2Al_S + AlCl_{3_G}.$$

This could be an attractive alternative to the combined Bayer process and Hall–Héroult process which is now used universally for the commercial production of aluminium.

Iron powder (99% pure) has been produced at Peace River, Canada,[28] by reduction of ferrous chloride with hydrogen at about 650°C,

$$FeCl_{2_S} + H_{2_G} = Fe_S + 2HCl_G,$$

in a process which could replace conventional iron-making in suitable regions. The ferrous chloride is produced by leaching partially reduced low grade iron oxide ores with hydrochloric acid followed by evaporation of the aqueous solution. The hydrochloric acid is regenerated in the reduction process.

Finally, and commercially this is the most important use of halides in extraction metallurgy, halides in general form highly conducting melts with low melting points. Oxides melt at high temperatures, and often tend to be poor electrolytic conductors, so that if a reactive metal is to be extracted by fused salt electrolysis, the electrolyte used is a halide rather than an oxide. Apart from improved conductivity, the lower melting point allows the process to be carried out at a lower temperature, which is always an advantage; also, if the temperature were too high, volatile metals such as magnesium would appear inconveniently as highly reactive gases when discharged at the cathode. Eighty per cent of the world's magnesium is extracted by electrolysis of $MgCl_2$ dissolved in a liquid $CaCl_2$, KCl, NaCl mixture. All aluminium production is by electrolysis of an electrolyte consisting of about 5% Al_2O_3 dissolved in a mixture of NaF, AlF_3 and CaF_2. MgO melts at 2800°C and Al_2O_3 at 2030°C, and both processes using molten halides are carried out below 1000°C. In addition, Al_2O_3 is a "network-forming" oxide like silica, and consequently would be a poor conductor (Section 5.2), so that the advantage of halides should be clear.

7.7. Refining by Preferential Oxidation

Usually, metals will be impure when they are initially extracted from their ores because the conditions under which they are extracted will also favour the extraction of impurity elements present as gangue minerals and these impurity elements will dissolve in the bulk metal. It will therefore be necessary to *refine* the crude extract to produce a purer metal. This metal may then be incorporated in an alloy, for example in steelmaking alloying elements such as manganese, nickel and chromium may be added after the refining process has been completed. Figure 7.2 demonstrates that different oxides have different stabilities, and this means that the equilibrium constant for the reactions by which they are produced will vary. If oxygen is introduced into a liquid alloy containing two metals, the metal forming the more stable oxide will be oxidized preferentially—that is the equilibrium constant for the reaction to form its oxide will be higher than that for the other oxide. This principle is used in the *fire-refining* of blister copper in which oxygen is introduced into the liquid copper by blowing air through pipes immersed in the metal. Elements forming more stable oxides than copper (Fe, Zn, S) will be oxidized preferentially, and will either form a slag or leave the metal as gases. For example, at 1300°C

$$2Fe + O_2 = 2FeO, \qquad \Delta G^\circ = -322 \text{ kJ}.$$

$$4Cu + O_2 = 2Cu_2O, \qquad \Delta G^\circ = -134 \text{ kJ}.$$

The Cu–Fe solution shows a positive deviation from Raoult's law (Section 3.3) so that Fe will have a high activity in the solution, and the FeO will be dissolved at low activity in a siliceous slag; this means that ΔG for the oxidation of iron under these conditions is likely to be less than −322 kJ. In the *softening* of lead, lead bullion from the blast furnace is oxidized by adding lead oxide. Impurities such as tin, arsenic and antimony are removed by preferential oxidation. (In a similar process, but using chlorine instead of oxygen, zinc can be removed from liquid lead by preferential chlorination to produce zinc chloride, which is more stable than lead chloride.)

Pig iron, the product of the iron blast furnace, contains about 4% carbon (dissolved from the coke), and elements whose oxides are reduced in the blast furnace—including up to 2% phosphorus, 2·5% silicon, 0·1% sulphur and 2·5% manganese—all the quantities expressed as % by weight. We will examine the *steelmaking* process, in which pig iron is refined, in some detail because it illustrates the principles of preferential oxidation and the use and control of slag/metal reactions.[29]

Oxygen may be introduced into the liquid pig iron in several ways—for example, by blowing air through the bottom of the vessel (Bessemer Converter process), blowing oxygen through steel pipes immersed in the metal (Rotor process and certain modifications of the Open Hearth and Electric Arc processes), blowing oxygen at high pressure on to the surface of the metal (L.D. process) and at slightly lower pressure which does not leave the surface of the metal bare by parting the slag cover (KALDO process, Rotor process, modifications of the Open Hearth process), and from an oxidizing slag (Open Hearth and Electric Arc processes may depend on this entirely but the other processes also use oxidizing slags). The KALDO and Rotor processes provide an additional stirring action by having a rotating vessel. Where almost pure oxygen is used instead of air the rate of the reactions should increase because of the increased concentration of one of the reactants (Section 4.8), and direct contact between the liquid metal and gaseous oxygen should give more rapid reactions because the slow step of mass transport in a slag layer will be cut out. Rates of reaction will depend on mass transport to and from gas/metal, slag/metal and gas/slag interfaces (Section 6.6).

Thermodynamically, carbon will be preferentially oxidized quite easily in steelmaking as ΔG° for

$$2C + O_2 = 2CO$$

is -552 kJ at 1600°C (the temperature at which we will assume steelmaking reactions occur for the purpose of this section). Carbon and oxygen will be dissolved in liquid iron at other than unit activity and p_{CO} may not be at 1 atm pressure, but ΔG will still have a large negative value if calculated from the van't Hoff isotherm (2.37). The problem of carbon removal in steelmaking is provision of a gas/metal interface at which the reaction

$$[C]+[O]=CO_G$$

can take place. Homogeneous nucleation of gas bubbles in liquid metals is impossible (Section 6.2) so that processes in which air or oxygen is introduced directly into the liquid bath should give rapid carbon removal (e.g. the Bessemer Converter process) compared with those dependent on nucleation of gas bubbles at cracks in the furnace lining (Open Hearth process). It is probably true that equilibrium is never reached in this reaction, but it will be approached more closely in the former than in the latter case. Carbon removal causes a mechanical stirring effect which is often vital to the successful completion of the steelmaking process and also its rate often controls the overall rate of the process and consequently the output of steel, so that this reaction is probably the most important in steelmaking.

Silicon and manganese removal usually present no problems because they form more stable oxides than iron (Fig. 7.2) and their oxides are liquid, less dense than liquid iron and consequently float out to enter the slag phase, whatever its composition.

Sulphur and phosphorus present an entirely different problem because their oxides are less stable than iron oxide at steelmaking temperatures. Blowing oxygen into liquid iron containing sulphur and phosphorus would mainly result in the oxidation of iron, and therefore there will be virtually no reaction in the bulk metal phase in steelmaking involving sulphur and phosphorus. Phosphorus will dissolve at very low activity in basic slags possessing excess oxygen ions,

$$2[P]+5[O]=(P_2O_5)$$

then $$(P_2O_5)+3(O^{2-})=2(PO_4)^{3-},$$

so that a basic slag is necessary for phosphorus removal in steelmaking. Acid steelmaking processes use acid slags in which all the oxygen supplied by the basic oxides is taken up by the silica (Section 7.2) and therefore acid processes must have a low phosphorus content in the charge to produce a commercially acceptable steel. The importance of taking lime into solution in the slag at an early stage to allow efficient phosphorus removal has prompted the injection of powdered lime with oxygen in the L.D.A.C. and O.L.P. processes, modifications of the original L.D. process. The sulphur transfer from metal to slag is by means of the reaction

$$[S]+(O^{2-})=(S^{2-})+[O].$$

This will be favoured by a basic slag (high free O^{2-} content), and a low oxygen activity in the metal, $a_{[O]}$. The basic slag can be provided, but all steelmaking processes depend on having oxidizing conditions so that they tend to be poor sulphur-removers. Only the Electric Arc process, which can replace the oxidizing slag by a reducing slag, is a good sulphur-remover, and the iron blast furnace is quite an efficient sulphur-remover because, although it uses an acid slag (Table 7.1), it works under reducing conditions with a high activity of sulphur in the metal compared with that in steelmaking. Because phosphorus and sulphur removal only take place at the slag/metal interface, good stirring and a large slag/metal interfacial area coupled with a fluid slag are vital for rapid transfer, hence the importance of the stirring action of the "carbon boil".

The formation of oxides in steelmaking is an exothermic process, and may be the only source of heat—as in the L.D., Bessemer, KALDO and Rotor processes. Silicon and phosphorus reactions are particularly important in this respect,

$$[Si] + O_2 = SiO_2, \qquad \Delta H_{1873} = -795 \text{ kJ}.$$

$$2[P] + \tfrac{5}{2}O_2 = P_2O_5, \qquad \Delta H_{1873} = -1213 \text{ kJ}.$$

The more efficient the combustion reactions, the more scrap which can be incorporated and melted in the charge. CO is blown out of the vessel before it can react with oxygen in the atmosphere in the Bessemer and L.D. processes, but in the KALDO a lower oxygen pressure is used and

$$CO + \tfrac{1}{2}O_2 = CO_2, \qquad \Delta H_{1873} = -280 \text{ kJ},$$

takes place inside the vessel, so that a higher scrap charge can be used to supplement the liquid pig iron. In the Open Hearth and Electric Arc processes, the use of an external source of fuel allows up to 100% solid metal in the charge (pig iron and/or scrap).

The high proportion of the basis metal being refined in these processes means that the metal itself must be oxidized because the rate of reactions like

$$[Fe] + [O] = (FeO)$$

in steelmaking and

$$2[Cu] + [O] = (Cu_2O)$$

in fire-refining will be high, and some considerable loss of iron and copper to the slag is therefore to be expected. In addition, oxygen will be left in solution in both these liquid metals at the end of the process. This would impair the mechanical properties of the solid metal so that some form of *deoxidation* is necessary before the metal is cast into ingots. In steelmaking, elements like aluminium and silicon are added. They form more stable oxides than iron (Fig. 7.2), and these oxides separate from the liquid steel by floating out (Section 6.3), thus removing the bulk of the oxygen. Vacuum degassing uses the fact that, by lowering p_{CO}, carbon becomes a very efficient deoxidant (Section 2.11) so that oxygen may be removed from liquid steel in addition to hydrogen and nitrogen by subjecting the steel to pressures of the order of 1 mm Hg.[30] Copper is deoxidized after fire-refining by stirring the copper with green wood poles. The metal is covered with a layer of charcoal or coke and the volatilization of the constituents of the wood causes an efficient stirring action. Deoxidation is probably by a combination of carbon and the distillation products of the wood, and reduces the oxygen content of the copper from 0·9 to about 0·04%. Further deoxidation may be by means of phosphor–copper additions,

$$5[O] + 2P = P_2O_5,$$

or by remelting and casting in a neutral atmosphere or a vacuum using carbon as a deoxidant if phosphorus cannot be tolerated in copper used for its high electrical conductivity.

7.8. Hydrometallurgical Processes

Minerals can, under certain circumstances, be taken into solution in water and the solution may be processed to effect a purification or to produce the metal directly. Clearly, not all minerals are soluble in water, but pretreatment can convert certain minerals into a soluble form, for example a sulphatizing roast can make sulphides water-soluble. Additions of acid, alkali or certain salts to the water can improve its solvent properties for certain minerals. To isolate the metal from the solution, the following methods may be employed:

Electrolysis (to be discussed in the next section).

Precipitation as an insoluble compound, for example by hydrolysis.

Reduction with hydrogen or a less noble metal which can cause precipitation of the metal from the solution.

Ion exchange and *solvent extraction*.

The first stage in the process is *leaching* of ores, concentrates or intermediate solid products (such as slags or precipitates) to take the value metal into aqueous solution. The advantages of leaching are that low grade ores or concentrates can be treated to produce a high yield of value metal where complex mineral dressing followed by pyrometallurgical processes might be costly and produce a low yield—for example, in gold extraction; very little fuel, other than for pumps, agitators, etc., is required; equipment is simple and inexpensive; an aqueous solution of a metal may be ideal for the next process—for example, electrolytic extraction from aqueous solution. Leaching is a heterogeneous process, involving three reaction steps:(31)

I. Diffusion of solvent through the pores in the solid particle and the dissolved substance outwards after solution.

II. Transport of the dissolved substance in the solvent away from the particle surface.

III. The chemical process of solution at the reaction site in the particle.

III is likely to be too fast to be rate-controlling, so that I and II must be made as rapid as possible by using a fine particle size to reduce the diffusion path and *agitation* to cause forced convection and prevent sedimentation forming a packed bed of fine particles. In some processes involving porous ores, agitation can be dispensed with and leaching is by *percolation*.(32) Worked-out deposits have been leached *in situ* by this technique to yield economic quantities of metal. Increasing the temperature usually increases solubility, and will increase diffusion rates in step I, but may be uneconomic in its consumption of fuel.

The solvent used should possess certain properties: it should be cheap and readily available—for example, the use of water in percolation leaching of sulphides (in which dissolved oxygen plays an important part) and the leaching of sulphates produced by roasting of sulphide concentrates.

Reactions with oxygen will take place, followed by solution, in the case of sulphide leaching,

$$Cu_2S + 2O_2 = 2CuO + SO_2$$

$$SO_2 + H_2O + \tfrac{1}{2}O_2 = H_2SO_4$$

$$CuO + H_2SO_4 = CuSO_4 + H_2O$$

$$CuSO_{4_S} = Cu^{2+}_{aq} + SO^{2-}_{4aq}.$$

Regeneration of the solvent helps to reduce costs; for example, where sulphuric acid solution is used for leaching oxidized copper ores and concentrates or zinc calcines prior to aqueous electrolysis, the sulphuric acid will be regenerated in the electrolytic process (Section 7.9). If possible, the solvent should not react with gangue minerals, and if a lime-rich gangue is present leaching may be by ammoniacal solutions, rather than acids; for example, copper will be taken into solution as the amine complex $[Cu(NH_3)_4]^{2+}$. The ammonia can be recovered and the copper reprecipitated if the ammonia is driven off by heating the clear solution. Finally, if possible, the process should be specific – gold-bearing ores are leached with cyanide solution in the presence of oxygen to dissolve the gold as a complex anion, $[Au(CN)_2]^-$. Lime additions are made to raise the pH of the solution to reduce the amounts of the more easily hydrolysed elements such as copper and iron entering the solution.

Pressure leaching at elevated temperatures has been used to increase solution rates and to dissolve minerals which may be less soluble at room temperature and atmospheric pressure. Forward and Mackiw[33] have described the principles of the process used at Sherritt Gordon Mines, Canada, in which a nickel-bearing sulphide concentrate containing cobalt, copper and iron is leached at about 80°C and 7·5 atm pressure with ammonia solution in the presence of oxygen. The iron sulphide is oxidized and remains undissolved as the hydrated oxide, but the nickel, copper and cobalt sulphides are taken into solution as the complex ammines, $[Ni(NH_3)_6]^{2+}$, etc. This solution is used in the subsequent purification process which is followed by reduction with hydrogen, again under pressure. In the Bayer process for the purification of alumina prior to fused salt electrolytic extraction of aluminium, alumina is dissolved from impure bauxite ore in 30% caustic soda solution at 160°C and 8 atm pressure. Under these conditions, the important gangue minerals (oxides of iron, silicon and titanium) are mainly undissolved, whereas the aluminate is soluble

$$Al_2O_3 + 2NaOH = 2NaAlO_2 + H_2O.$$

After removal of the residue, the alumina is reprecipitated as the hydrate, Al_2O_3, $3H_2O$ by allowing the solution to cool in the presence of seeding crystals of the hydrate. Calcination of the precipitate completes the process (Section 7.4).

Isolation of cobalt from purified leach solution at Rhokana, Zambia, is as an insoluble hydroxide as a result of hydrolysis in a solution of pH 8·3. The precipitate is redissolved in sulphuric acid to form an essentially pure electrolyte for aqueous electrolytic extraction.[34] Another form of isolation is by *gaseous reduction* under pressure at elevated temperatures, whose principles are reviewed in a paper by Meddings and Mackiw.[35] This process has been used at Sherritt Gordon Mines as an integral part of the important development of the pressure leaching process described above. Carbon monoxide and sulphur dioxide have been used for gaseous reduction but hydrogen is most widely used and will be considered here. The overall reaction for the reduction of a divalent metal ion M^{2+} by hydrogen in aqueous solution is

$$M^{2+} + H_2 = M + 2H^+, \qquad (7.19)$$

which is a direct replacement of the metal ions in solution by hydrogen ions. In Section 5.6, it was explained that the tendency of an element to leave an electrolyte to discharge at an electrode could be expressed by the electrode potential.

Equation (7.19) can be considered thermodynamically as if it takes place in two separate reaction steps,

$$M^{2+} + 2\epsilon^- = M \tag{7.20}$$

and

$$H_2 = 2H^+ + 2\epsilon^-. \tag{7.21}$$

Equation (7.20) can be considered as the electrode reaction taking place at the metal electrode M in contact with a solution of its ions and (7.21) represents a hydrogen electrode. From (5.23), the electrode potential of (7.20) is given by

$$E_M = E_M^0 + \frac{RT}{2F} \ln a_{M^{2+}} = E_M^0 + \frac{2{\cdot}303RT}{2F} \log a_{M^{2+}}, \tag{7.22}$$

and the electrode potential of the hydrogen electrode, by (5.29), is

$$\begin{aligned} E_H &= E_H^0 + \frac{RT}{F} \ln a_{H^+} \\ &= E_H^0 + \frac{2{\cdot}303RT}{F} \log_{10} a_{H^+} \\ &= E_H^0 - \frac{2{\cdot}303RT}{F} \text{pH}. \end{aligned} \tag{7.23}$$

The lower the electrode potential, the greater the tendency of the metal or hydrogen to dissolve in the electrolyte, so that reductions will occur by reaction (7.19) when $E_H < E_M$. Equation (7.23) assumes that the hydrogen gas is at 1 atm pressure, but if p_{H_2} varies, it will be included in (7.23) as follows:

$$\begin{aligned} E_H &= E_H^0 + \frac{RT}{F} \ln \frac{a_{H^+}}{p_{H_2}^{1/2}} \\ &= E_H^0 - \frac{2{\cdot}303RT}{F} \text{pH} - \frac{2{\cdot}303RT}{2F} \cdot \log_{10} p_{H_2}. \end{aligned} \tag{7.24}$$

E_H and E_M are plotted in Fig. 7.6 (after Meddings and Mackiw[35]) as functions of pH of the solution. E_M is independent of pH, and is plotted for various metal electrodes, and E_H is plotted for two values of p_{H_2}. The temperature and the activity of the metal ions in the electrolyte are assumed to be constant throughout. If at any point on the diagram $E_H <$ E_M, reduction of the metal ion will occur. Thus, at any pH, hydrogen will replace copper from a molar solution of copper ions, but a pH of at least about 4 for nickel and 9 for iron would be necessary. The effect of increasing p_{H_2} is to lower the pH

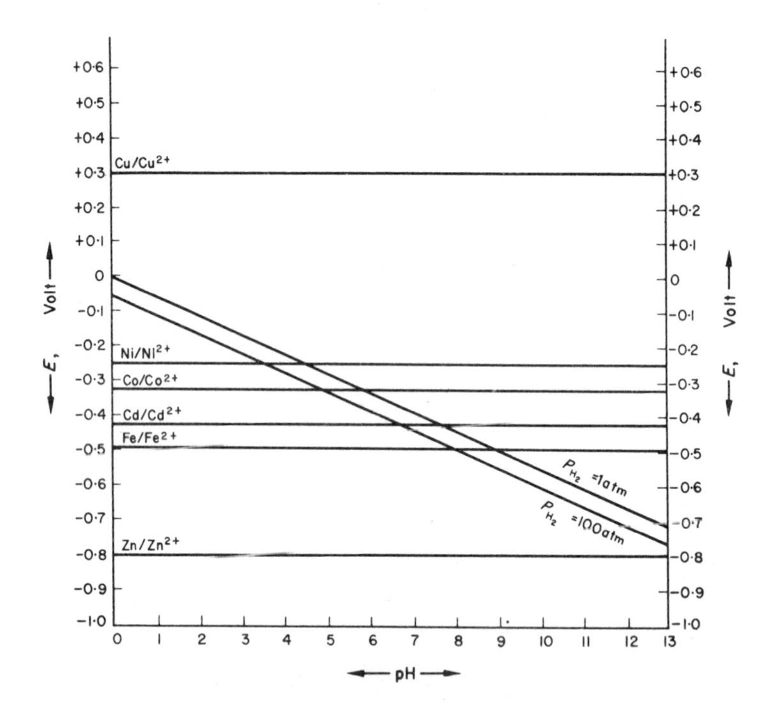

FIG. 7.6. E_M and E_H plotted as a function of pH for electrodes $M|M^{2+}_{aq}$ and $H_2|H^+_{aq}$ at 25°C with a molar solution of M^{2+}. Two lines for E_H are shown for $p_{H_2} = 1$ atm and 100 atm. (After Meddings, B. and Mackiw, V. N.[35])

at which reduction of a metal ion occurs. The hydrogen overpotential and metal overpotentials are ignored in this argument and it must be assumed that the high hydrogen overpotential on zinc would have a considerable influence on reduction of Zn^{2+} by hydrogen. We thus have a method of *selective* reduction of metals from a mixed solution. Reduction at pH2 would precipitate copper powder from a solution containing Cu^{2+}, Ni^{2+} and Co^{2+}. The pH could then be increased and the nickel and cobalt would be reduced. If the pH is greater than 7, hydrolysis tends to occur, for example $Ni(OH)_2$ would be precipitated. This can be prevented by using ammonia to raise the pH and to form complex ions with the metals present, $[Ni(NH_3)_6]^{2+}$, etc. This will lower $a_{Ni^{2+}}$ and the position of the nickel line on Fig. 7.6. The rate of the reaction will be increased at elevated temperature and pressure – not only due to the rise in temperature but also the increased solubility of the hydrogen gas in water at high pressures will raise the concentration of one of the reactants. Therefore the reduction of nickel is carried out at about 170°C and 30 atm pressure. The kinetics of nucleation of the metal particles are mentioned briefly in Section 6.5.

Cementation is an older process than pressure reduction, but is also based on electrochemical principles. Referring to Fig. 7.6, if iron metal is introduced into a solution of copper ions, $E_{Fe} < E_{Cu}$, so that the iron will displace the copper from the solution. Scrap iron has been used to recover copper from leach solutions, nickel to remove copper ions from electrolytes used for electrolytic refining of nickel, and cobalt granules to remove copper from cobalt electrolytes. Zinc dust is added to displace gold from cyanide leach solutions. (See also Section 5.9.)

In *ion exchange* processes, metal ions are preferentially adsorbed at active sites on an otherwise inert resin when aqueous solutions are passed through a column containing the resin. The adsorbed ion can then be removed by "elution" – a strong solution of other ions is passed through the column and the value metal ions are displaced from the resin. In *solvent extraction,* metal ions in aqueous solution are complexed by an organic reagent and the complex is dissolved in an organic solvent such as paraffin, which is immiscible with water. At equilibrium, the partial molar free energy and hence the activity of the metal ions will be the same in both water and solvent phases (Chapter 3), but the activity coefficient of the metal ions in the complex will be very much lower than in the aqueous solution. Hence the *partition coefficient,* which is the ratio of the concentration of the metal ions in the solvent phase to the concentration of metal ions in the water, will be high and the metal ions will be extracted from the aqueous solution. Both ion exchange and solvent extraction can be specific to certain ions, and have been used to isolate uranium ions from very dilute leach solutions[36] and to separate metals such as niobium and tantalum which are otherwise chemically similar and hence difficult to separate.

7.9. Electrolysis

Electrometallurgical processes include extraction of metals from solution and refining metals by electrolysis using aqueous and fused salt electrolytes. Electrochemistry was discussed in Chapter 5 and electrodeposition will also be considered in Chapter 8 so that in this section we will indicate some of the factors affecting the choice of electrolysis as an extraction or refining process and the control of the process.

The cathode reaction in all the processes can be represented as

$$M^{z+} + Z\epsilon^{-} = M, \tag{5.15}$$

but the anode reaction will depend on the electrolyte used and on whether the process is extraction or refining. In *extraction,* an anode is used which does not dissolve electrolytically, such as lead in copper extraction from aqueous solution or graphite in the extraction of magnesium from a fused chloride electrolyte. The anode reaction in aqueous solution will be

$$2(OH^-) = H_2O + \tfrac{1}{2}O_2 + 2\epsilon^-, \qquad (7.25)$$

in a fused chloride electrolyte,

$$Cl^- = \tfrac{1}{2}Cl_2 + \epsilon^- \qquad (7.26)$$

and in a fused oxide electrolyte

$$O^{2-} = \tfrac{1}{2}O_2 + 2\epsilon^-. \qquad (7.27)$$

Although the electrolyte used in aluminium extraction consists primarily of fluorides, aluminium oxide is dissolved in it and the ion discharged at the anode is oxygen rather than fluorine. In *refining* processes an impure metal anode is used, and metal ions enter the electrolyte by the reverse of reaction (5.15),

$$M = M^{z+} + z\epsilon^-. \qquad (7.28)$$

The processes are run so that the potential at the cathode is sufficient to discharge the metal ions required and in refining at the anode to dissolve the necessary metal ions. Ions will discharge at the cathode when its potential is less than $E_M + \eta_c$, that is the discharge potential, the sum of the reversible electrode potential and the cathode overpotential (Section 5.13), and all ions present in the electrolyte for which the cathode potential is less than their discharge potentials will discharge at the cathode. If a metal ion has a discharge potential less than the hydrogen discharge potential (this includes hydrogen overvoltage), then it cannot be extracted by electrolysis from aqueous solution because hydrogen will discharge at the cathode rather than the metal required. This is the reason for the choice of fused salt electrolytes for the extraction of magnesium and the extraction and refining of aluminium (see Table 5.1, p. 196, for a comparison of standard electrode potentials). Zinc can be discharged from aqueous solution because of the high hydrogen overvoltage on zinc, but small quantities of impurities in the electrolyte—especially cobalt combined with germanium—lower the hydrogen overvoltage on zinc and cause a loss in current efficiency because hydrogen discharges at the cathode. Careful purification of the zinc sulphate electrolyte is necessary in extraction from aqueous leach solutions, but this results in a very pure product (99·99% pure zinc) which does not need refining by fractional distillation as is the case with zinc produced by smelting and used in certain commercial applications requiring high purity.[37] Half the world's zinc is produced by aqueous electrolysis.

In extraction processes, the e.m.f. applied is given by (5.40),

$$V = E_{\text{cell}} + |\eta_c| - |\eta_a| - IR,$$

where η_c and η_a are the overpotential at the cathode and anode respectively, I the current, R the resistance in the circuit. E_{cell} is the e.m.f. required to drive the overall cell reaction against a chemical free energy change which tends to oppose the electrolysis process. For example, for the extraction of magnesium at 750°C from a fused chloride electrolyte, the overall cell reaction is

$$MgCl_2 = Mg + Cl_2, \qquad \Delta G^\circ = +460 \text{ kJ}.$$

Using (5.22),

$$\Delta G^\circ = -E_{\text{cell}} zF = -2E(96{,}500).$$

E is measured in volts and ΔG° is in kJ so that

$$E_{\text{cell}} = -\frac{460{,}000}{96{,}500 \times 2} = -2{\cdot}38 \text{ V}.$$

Overpotential and circuit resistance raise the applied voltage required to about −7 V, but with a current of 20,000 A, the power consumption will be high. *Refining* processes require very little power to overcome E_{cell} because the two electrodes have the metal present at almost the same activity—the cathode consists of the pure metal, so that $a_{M^{z+}} = 1$, and the anode is often at least 98% pure so that the value of E_{cell} will be that of a concentration cell, given by (5.35),

$$E_{cell} = \frac{RT}{zF} \ln \frac{(a \text{ of } M^{z+} \text{ in cathode})}{(a \text{ of } M^{z+} \text{ in anode})} \simeq 0.$$

(Note that in this case it is the activity of the metal in the *electrode* which differs, rather than its activity in the electrolyte surrounding the electrode.) Aluminium is refined by having a liquid cathode of pure aluminium (sp.gr. 2·3) which floats on the electrolyte (Na_3AlF_6, $BaCl_2$, sp.gr. 2·8). The electrolyte floats on the liquid anode (an aluminium—30% copper alloy, sp.gr. 4·5). E_{cell} is merely a few millivolts, but the applied voltage is raised to − 1·5V by overpotential and resistance may add several more volts. Impure aluminium and scrap are added to the liquid anode and pure aluminium (99·99%) can be removed from the cathode. An interesting feature of the Hall-Héroult process for the *extraction* of aluminium is that the overall cell reaction at 970°C is

$$\frac{2}{3}Al_2O_3 = \frac{4}{3}Al + 2O, \qquad \Delta G^{\circ} = +837 \text{ kJ} \qquad \text{(from Fig. 7.2)}.$$

But the oxygen evolved at the carbon anode reacts with carbon to form carbon dioxide (the kinetics of this reaction seem to be favourable compared with those for the formation of carbon monoxide, which is thermodynamically more stable in the presence of carbon at this temperature).

$$C + 2O = CO_2, \qquad \Delta G^{\circ} = -402 \text{ kJ} \qquad \text{(Fig. 7.2)}.$$

The overall reaction is therefore the sum of the cell reaction and this latter reaction,

$$\frac{2}{3}Al_2O_3 + C = \frac{4}{3}Al + CO_2, \qquad \Delta G^o = +837 - 402 = +435 \text{ kJ}.$$

Carbon is in effect acting as a reducing agent, lowering the voltage necessary to supply electrical energy to dissociate the alumina.

$$E_{cell} = -\frac{435{,}000}{96{,}500 \times 4} \qquad \text{(from 5.22)}$$

$$= -1{\cdot}13 \text{ V}.$$

This is raised to about −5 V to overcome polarization and resistance. The current used is up to 100,000 A.[38]

Power consumption tends to be much lower in aqueous electrolysis, but the same principles apply as in the case of fused salt electrolysis. The example of copper refining in aqueous solution, using a copper sulphate–sulphuric acid electrolyte, demonstrates an interesting point about the refining action which takes place at both the anode and cathode as follows: considering only standard electrode potentials (account should, of course, be taken of activities of metal ions in the electrolyte and metals in the electrode but the argument remains the same), and a copper anode containing only silver and iron as impurities, from Table 5.1, $E^o_{Cu} = +0{\cdot}34$ V, $E^o_{Ag} = +0{\cdot}80$ V, $E^o_{Fe} = -0{\cdot}44$ V. If the cell is run so that copper dissolves at the anode, the iron will also dissolve but the silver will not, so that it remains on the anode and may eventually be recovered from the "anode slime". At the cathode, the potential will be such that the copper will discharge but the iron, being of lower discharge potential, will not. Thus preferential dissolution at the anode and preferential discharge at the cathode ensure that the copper is purified. In practice, electrode potentials are measurements of the position of *equilibrium* in electrode reactions so that very small amounts of impurities will tend to discharge with the copper, and as these build up in the electrolyte this tendency increases so that the electrolyte must be replaced continuously. In extraction, the electrolyte will become impoverished in the value metal, so that it must be replenished continuously. This is convenient in the case of copper where the electrolyte has been produced by leaching concentrates or ores with sulphuric acid. During electrolysis, reaction (5.15) causes a depletion in Cu^{2+}, and (7.25) a depletion in OH^- and consequently an increase in H^+ because

$$H_2O = H^+ + OH^-$$

must take place to produce more OH^- ions. SO_4^{2-} ions remain in solution so that the H_2SO_4 concentration builds up until, after purification, the spent electrolyte can be used once again as the leaching solvent. The advantage of electrolytic refining of copper is that elements like Ag, which form less stable oxides than copper and cannot be removed in fire-refining (Section 7.7), can be removed and, in this case, recovered as a valuable by-product from the anode slime. Bismuth and silver can be removed from lead by aqueous electrolysis but not by softening for the same reasons.

Choice and control of the electrolyte is important. Apart from the choice between aqueous and fused salt electrolytes, the choice of aqueous electrolyte for lead, for example, is controlled by the fact that lead forms an insoluble sulphate and a peroxide of lead is produced in solutions of other mineral acids. An electrolyte of lead fluosilicate in hydrofluosilicic acid (H_2SiF_6) is used to avoid these problems. Acids are added to aqueous electrolytes to lower their resistance without adding ions which will discharge at the cathode—although too high an acid concentration would cause a hydrogen discharge. Pure fused salts are never used in fused salt electrolysis, and a mixture is used to lower the melting point and increase the conductivity. Transport number measurements indicate that most of the current is carried by sodium ions rather than aluminium ions in the Al_2O_3–NaF–AlF_3 electrolyte in the extraction of aluminium. Good stirring of the electrolyte is essential in all processes to prevent local concentration build-up, and the temperature in aqueous electrolytes may be increased to about 50°C to lower the electrolyte resistance without causing loss due to evaporation and an unpleasant atmosphere. The electrolyte in fused salt electrolysis is kept molten by the heat produced by the passage of the electric current.

In nickel refining, using an aqueous electrolyte of $NiCl_2$ and $NiSO_4$, the electrolyte is removed from the anode compartment before it comes into contact with the cathode. This is to prevent discharge of Fe^{2+}, Co^{2+} and Cu^{2+}, which dissolve at the anode to some extent and would discharge at the cathode at the potential used to deposit the nickel. The electrolyte is purified before being led into the cathode compartment, which is kept at a slightly higher liquid level so that liquid flow is if anything away from the cathode through the canvas diaphragm separating the two compartments. (This process has been extended by the International Nickel Company to allow refining of nickel directly from anodes cast from the nickel sulphide concentrate left after the separation by froth flotation from Bessemer matte—Section 7.5. The anode reaction is

$$Ni_3S_2 = 3Ni^{2+} + 2S + 6\epsilon^-.)^{(39)}$$

Finally, a list of metals which may be electrolytically extracted or refined should emphasize the importance of electrometallurgy in the field of extraction metallurgy:[2]

Aqueous electrolysis: Sb, Bi, Cd, Co, Au, Ag, Mn, Sn, Cr, Cu, Ni, Pb, Zn.

Fused salt electrolysis: Al, Mg, Be, Na, Ca, Ce, Nb, Ta.

References

1. International Nickel Company of Canada Ltd. *Milling and Smelting the Sudbury Nickel Ores.* (Film, 16 mm, Sound, Colour.)
2. DENNIS, W. H. *Metallurgy of the Non-ferrous Metals*, Pitman, London, 2nd edn., 1961: BRAY, J. L. *Ferrous Process Metallurgy*, Wiley, New York, 1954: BRAY, J. L. *Non-ferrous Production Metallurgy*, Wiley, New York, 2nd edn., 1947: ELLIOTT, G. D. and BOND, J. A. *Practical Ironmaking*, The United Steel Companies Ltd., Sheffield, 1959: BASHFORTH, G. R. *The Manufacture of Iron and Steel*, Vols. I and II, Chapman & Hall, London, 3rd edn., 1964: CHATER, W. J. B. and HARRISON, J. L. *Recent Advances with Oxygen in Iron and Steelmaking*, Butterworths, London, 1964.
3. CLARKE, F. W. and WASHINGTON, H. S. U.S. Geol. Survey Profess. Paper 127, 1924.
4. PARK, C. F. Jr., and MACDIARMID, R. A. *Ore Deposits*, W. H. FREEMAN, San Francisco, 1964.
5. GAUDIN, A. M. *Principles of Mineral Dressing*, McGraw-Hill, New York, 1939: PRYOR, E. J. *Mineral Processing*, Elsevier, Amsterdam, 3rd edn., 1965.
6. WARD, R. G. *An Introduction to the Physical Chemistry of Iron and Steel Making*, Edward Arnold, London, 1962, p. 16.
7. SCHENCK, H. *The Physical Chemistry of Steelmaking*, Springer, Berlin, 1932. English Translation, B.I.S.R.A., London, 1945.
8. TEMKIN, M. *Acta Physicochimica U.R.S.S.* **20** (1945), 411.
9. FLOOD, H. and GRJOTHEIM, K. *J.I.S.I.* **171** (1952), 64: FLOOD, H., FØRLAND, T. and GRJOTHEIM, K. *The Physical Chemistry of Melts*, Inst. Min. Met., London, 1953, p. 46.
10. ELLINGHAM, H. J. T. *J. Soc. Chem. Ind.* **63** (1944), 125.
11. RICHARDSON, F. D. and JEFFES, J. H. E. *J.I.S.I.* **160** (1948), 261, and *J.I.S.I.* **171** (1952), 167: HOPKINS, D. W. *Physical Chemistry and Metal Extraction*, Garnet Miller, London, 1954, p. 91 ff.: KELLOGG, H. H. *Trans. A.I.M.M.E.* **188** (1950), 862, and *Trans. A.I.M.M.E.* **191** (1951), 137.
12. MATHEWSON, C. H. (Ed.) *Zinc, The Metal, its Alloys and Compounds*, Reinhold, New York, 1959. Chapters 4 and 6: ref. 11, HOPKINS, D. W. Chapter IX.
13. PIDGEON, L. M. *Trans. Can. Inst. Min. Met.* **49** (1946), 621: PIDGEON, L. M. and KING, J. A. *The Physical Chemistry of Process Metallurgy*, *Disc. Faraday Soc.*, No. 4, 1948, Butterworths, London, p. 197.
14. MORGAN, S. W. K. *Trans. Inst. Min. Met.* **66** (1956–7), 553: MORGAN, S. W. K. and LUMSDEN, J. *J. Metals* (April 1959), 270: WOODS, S. E. and TEMPLE, D. A. *Trans. Inst. Min. Met.* **74** (1964–5), 297.
15. KIRKALDY, J. S., and WARD, R. G. (Eds.) *Aspects of Modern Ferrous Metallurgy*, Blackie, London, 1964. Contribution by ROSS, H. U., p. 65; ref. 6, Chapter 17.
16. BODSWORTH, C. *Physical Chemistry of Iron and Steel Manufacture*, Longmans, London, 1963, Chapter 6.

17. FORNANDER, S. *J.I.S.I.* **177** (1954), 7.
18. KELLOGG, H. H. and BASU, S. K. *Trans. Met. Soc. A.I.M.E.* **218** (1960), 70.
19. QUENEAU, P. (Ed.) *Extractive Metallurgy of Copper, Nickel and Cobalt*, Interscience, New York, 1961. Paper by THOMPSON and ROESNER, p. 3.
20. Ref. 12, pp. 136–63.
21. MCBRIAR, E. M., JOHNSON, W., ANDREWS, K. W. and DAVIES, W. *J.I.S.I.* **177** (1954), 316.
22. OLT, T. F. *J.I.S.I.* **200** (1962), 87: KNEPPER, W. A. (Ed.) *Agglomeration*, Interscience, New York, 1962: YOUNG, P. A. *Iron and Steel* (Sept. 1965), 455.
23. RUDDLE, R. W. *The Physical Chemistry of Copper Smelting*, Inst. Min. Met., London, 1953.
24. International Nickel Co. *Trans. Can. Inst. Min. Met.* **58** (1955), 158: BRYK, P. *et al., J. Metals*, **10** (1958), 395: SADDINGTON, R., CURLOOK, W. and QUENEAU, P. *J. Metals* (April 1966), p. 440.
25. Ref. 19, paper by SPROULE, K., HARCOURT, G. A. and RENZONI, L. S., p. 33.
26. KROLL, W. J. *Trans. A.I.M.E.* **215** (1959), 546.
27. GROSS, P. *et al. The Physical Chemistry of Process Metallurgy, Disc. Faraday Soc.*, No. 4, 1948, Butterworths, London, p. 206.
28. GRAVENOR, C. P. *et al., Can. Min. Met. Bull* (1964), 421.
29. References 6 and 16 provide a good survey of the physico-chemical principles of steelmaking, and are comprehensive sources of reference on this subject.
30. BUNSHAH, R. F. (Ed.) *Vacuum Metallurgy*, Reinhold, New York, 1958: BELK, J. A. *Vacuum Techniques in Metallurgy*, Pergamon, London, 1963: Iron and Steel Institute, London, Special Report No. 92. *Vacuum Degassing*, 1965.
31. COULSON, J. M. and RICHARDSON, J. F. *Chemical Engineering*, **2**, Pergamon, London, 1956, Chapter 17.
32. WADSWORTH, M. E. and DAVIS, F. T. (Eds.) *Unit Processes in Hydrometallurgy* (Met. Soc. A.I.M.E.), Gordon & Breach, New York, 1964. Contribution by SEIDEL, D. C., p. 114.
33. FORWARD, F. A. and MACKIW, V. N. *J. Metals* (1955), 457.
34. *Cobalt Monograph*, Centre D'Information du Cobalt, Brussels, 1960, p. 37.
35. Ref. 32, Contribution by MEDDINGS, B. and MACKIW, V. N., p. 345.
36. WILKINSON, W. D. *Uranium Metallurgy*, **1**, *Uranium Process Metallurgy*, Interscience, New York, 1962.
37. Ref. 12, p. 174.
38. GERARD, G. and STROUP, P. T. (Eds.) *Extractive Metallurgy of Aluminium*, Vols. I and II, Met. Soc. A.I.M.E., 1963 (Interscience, New York).
39. RENZONI, L. S., MCQUIRE, R. C. and BARKER, W. V. *J. Metals*, **10** (1958), 414.

CHAPTER 8

Corrosion and Electrodeposition

8.1. Introduction

Extraction of metals from their ores is an unnatural process in which the components of a stable system are separated, and if those components are given the opportunity, they will return to their stable state. Corrosion can be defined as the reaction of a metal with its environment; water and oxygen are the most plentiful of the reactive substances in a natural environment so that it is to be expected that the oxides and hydrated oxides of metals will be the principal products of corrosion. Sulphur and carbon dioxide and, especially in marine atmospheres, chlorides, are also present, and sulphides, basic sulphates, carbonates, and chlorides are to be expected amongst corrosion products in addition to oxides. This chapter is not a survey of the corrosion field, for which there are several references[1-4] which the reader is advised to consult, but is an extension of some of the principles set out earlier in this book in a particular field of chemical metallurgy. Some of the factors affecting the tendency to corrode and the rate of corrosion and some principles of corrosion protection will be covered, ending with a short treatment of electrodeposition, which is a topic closely associated with corrosion.

8.2. Oxide Films

It will be seen from Fig. 7.2, the Ellingham diagram for oxides, that most metal oxides have negative standard free energies of formation at temperatures up to 2000°C – this means that a metal will tend to react with oxygen to form an oxide film in this temperature range. Important exceptions to this are silver and gold–silver oxide dissociates above about 250°C, at which temperature ΔG° for the formation of Ag_2O becomes positive, and the oxide of gold is unstable at room temperature – it consequently does not appear on Fig. 7.2 and, being unreactive, gold tends to occur naturally in the uncombined "native" form. The presence of oxide films on the surfaces of metals has an important effect on corrosion in wet conditions, but the factors controlling the type and growth of oxide films in "dry" conditions will be considered first. (The stability of oxides is discussed in Sections 2.11 and 7.3.)

Sometimes, more than one oxide of a metal exists and therefore the oxide formed will depend on the position of equilibrium in the reaction to form the metal oxide M_xO_y,

$$xM + \frac{y}{2}O_2 = M_xO_y,$$

relative to the positions of equilibrium in the reactions by which the other oxides of the metal are formed. An example of this is the formation of three oxides of iron, wüstite (FeO), magnetite (Fe_3O_4) and haematite (Fe_2O_3). In Section 5.2, it was explained that solid ionic salts do not have perfect crystal lattices and are often non-stoichiometric; they possess deficiencies in their cation or anion lattices, and this is the case with most metal oxides. The formulae given above for the oxides of iron are therefore inaccurate, and we expect structures corresponding to formulae such as $Fe_{1-\delta}O$, where δ will vary according to the temperature and the partial pressure of oxygen in contact with the oxide. In anion-deficient oxides, the deficiency will decrease as the oxygen partial pressure increases, whereas the deficiency increases with oxygen partial pressure in cation-deficient oxides. If an oxide is in equilibrium with the metal from which it has formed, the extent of non-stoichiometry will be fixed at constant temperature. Darken and Gurry[5] give diagrams (Fig. 8.1) in which the variable composition of the oxides and the transformation from one oxide to another as the partial pressure of oxygen increases are shown clearly. Figure 8.1b can be interpreted in a similar

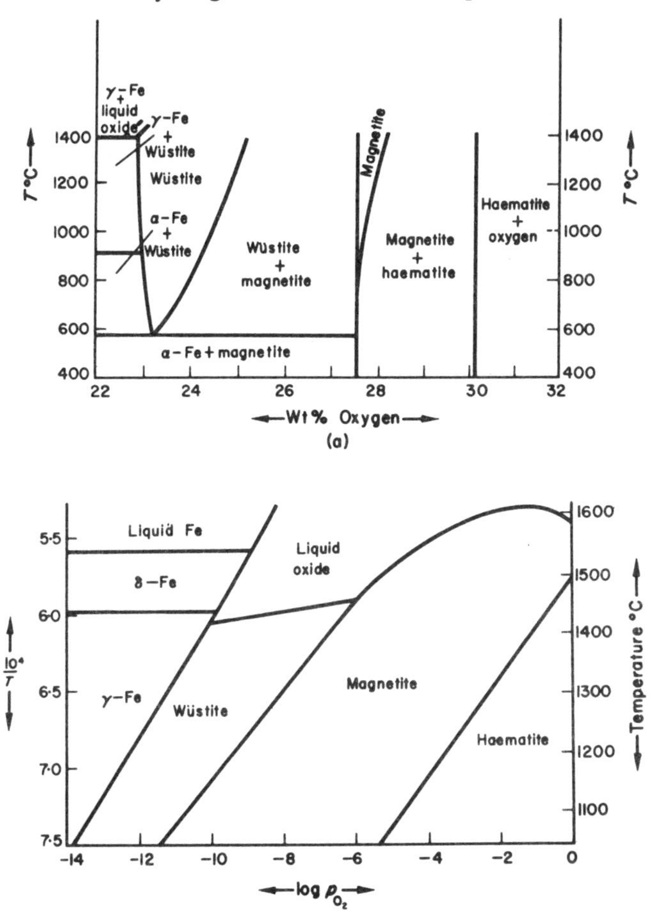

FIG. 8.1. Iron–oxygen system. (a) Temperature–composition diagram. (b) Temperature–pressure diagram for iron at unit activity. (After Darken, L. S. and Gurry, R. W.[5])

manner to Fig. 7.5 (Section 7.4), except that the variables are $10^4/T$ and log p_{O_2} instead of log p_{SO_2} and log p_{O_2}. From Fig. 8.1a, it will be seen that wüstite is unstable below 570°C, decomposing to form magnetite and $\alpha-$Fe,

$$4FeO = Fe_3O_4 + \alpha - Fe,$$

though equilibrium may be difficult to achieve for kinetic reasons. Iron heated above 570°C would be expected to form three layers—wüstite next to the metal where the oxygen activity is lowest, haematite on the outside where p_{O_2} is a maximum, and magnetite between the wüstite and the haematite. The extent of their non-stoichiometry would be expected to alter throughout the film so long as equilibrium was not achieved. Iron heated below 570°C would produce a scale consisting only of magnetite and haematite.

The kinetics of oxide film growth are no longer those of a heterogeneous reaction between oxygen gas and a solid metal surface, once the film is thicker than a monolayer. At very low temperatures, the film may be no more than a monolayer, but if a steel ingot is placed in an oxidizing atmosphere at high temperature, it may develop an oxide scale almost an inch thick. The gaseous oxygen will then be separated from the metal surface and mass transport through the oxide film can become the rate-controlling factor in the growth of the film. From Fig. 7.2, it can be assumed that metal oxides become generally less stable as the temperature increases so that the tendency to form oxides decreases with increase in temperature even though the rate of formation of the oxide film increases as a result of increased rates of reaction and mass transport, and changes in the film structure.

If the oxide film completely covers the surface of the metal, with no pores or cracks in the film allowing direct access of the gas to the metal, and with no discontinuities such as blisters to obstruct diffusion through the film, it is to be expected that as y, the thickness of the film, increases, the rate of increase of y will decrease, and it would be reasonable to write, for a diffusion-controlled process,

$$\frac{dy}{dt}=\frac{A}{y}, \tag{8.1}$$

where A is a constant depending on temperature and the nature of the film. Rewriting (8.1) and integrating,

$$y\,dy = A\,dt,$$

$$y^2 = 2At + B, \tag{8.2}$$

where B is the constant of integration. A graph of y plotted against t would be a parabola, and this known as the *parabolic "law" of oxidation.* Copper at 850°C, iron above 200°C and nickel at 900°C are known to obey a growth equation such as (8.2).

The oxides Cu_2O, NiO and FeO are cation-deficient (Fig. 8.2), and growth of nickel oxide, for example, will take place

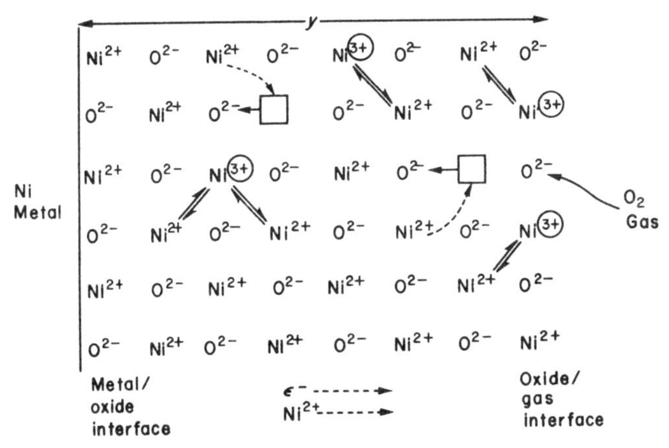

FIG. 8.2. Representation of the growth of the oxide film on the surface of nickel.

by the diffusion of Ni^{2+} ions via the vacant lattice sites in the Ni^{2+} lattice of the oxide outwards from the metal/oxide interface to the oxide/gas interface. Nickel oxide is a *p-type semiconductor* so that the electrons produced at the metal/oxide interface by the reaction

$$Ni = Ni^{2+} + 2\epsilon^-$$

are conducted to the oxide/gas interface by *positive-hole* conduction, a resonance between Ni^{2+} and Ni^{3+} which is transferred across the film as a result of the potential drop caused by the action of this chemical cell; the positive-hole conduction can be represented by the equation

$$Ni^{2+} \rightleftharpoons Ni^{3+} + \epsilon^-,$$

two Ni^{3+} ions being present to electrically balance the effect of each vacant site in the Ni^{2+} lattice. At the oxide/gas interface, the oxygen is adsorbed on the surface of the oxide film, where the reaction

$$\tfrac{1}{2}O_2 + 2\epsilon^- = O^{2-}$$

takes place. The growth of this type of oxide film can therefore be seen to involve two mechanisms, cationic diffusion and p-type electronic conduction. Zinc oxide at elevated temperatures is an *n-type semiconductor* with an anion-deficient structure, $Zn_{1+\delta}O$ (Fig. 8.3). The excess zinc ions diffuse via interstitial sites from the metal/oxide interface where the reaction

$$Zn = Zn^{2+} + 2\epsilon^-$$

takes place. The excess electrons then move through the oxide by *negative-type* conduction. At the oxide/gas interface, the oxygen ions are formed after the oxygen gas has

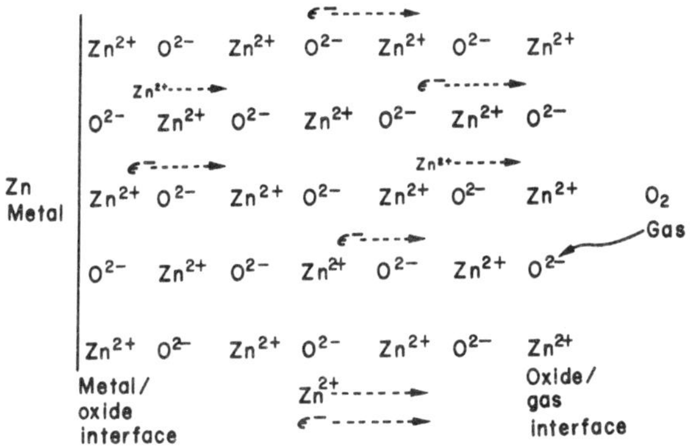

e^- ------→
Zn^{2+} O^{2-} Zn^{2+} O^{2-} Zn^{2+} O^{2-} Zn^{2+}
Zn^{2+} ----→ e^- ------→
O^{2-} Zn^{2+} O^{2-} Zn^{2+} O^{2-} Zn^{2+} O^{2-}
e^- ----→ Zn^{2+} ----→
Zn Metal Zn^{2+} O^{2-} Zn^{2+} O^{2-} Zn^{2+} O^{2-} Zn^{2+} O_2 Gas
e^- ----→
O^{2-} Zn^{2+} O^{2-} Zn^{2+} O^{2-} Zn^{2+} O^{2-}
Zn^{2+} O^{2-} Zn^{2+} O^{2-} Zn^{2+} O^{2-} Zn^{2+}
Metal/ oxide interface Zn^{2+} ------→ e^- ------→ Oxide/ gas interface

FIG. 8.3. Representation of the growth of the oxide film on the surface of zinc.

adsorbed on the oxide surface and each oxygen atom has picked up two electrons. If there are vacant sites on the anion lattice, growth will be by diffusion of oxygen ions inwards rather than outward metal ion diffusion, and this may be the case with zirconium, for which experimental evidence is not so clearly defined but whose oxide is an *n*-type semiconductor (Fig. 8.4). It is known that ferric oxide (Fe_2O_3) also grows by inward anion diffusion.

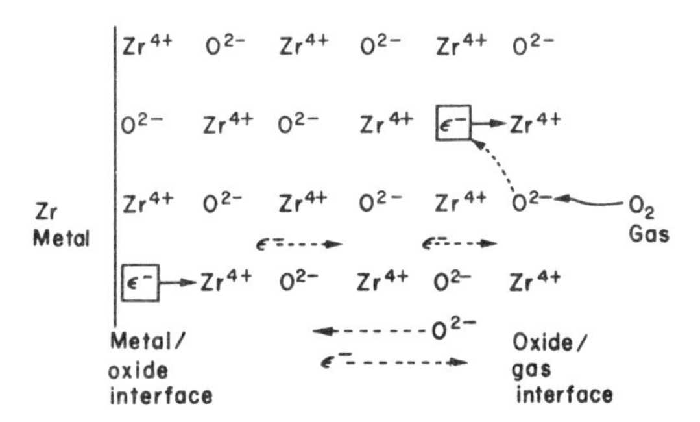

FIG. 8.4. Possible *n*-type semiconduction in the growth of the oxide film on zirconium. (Actual structure corresponds to the formula $ZrO_{2-\delta}$.)

Rapid oxide film growth will therefore depend on good ion conduction and good electron conduction, and Wagner, followed by Hoar and Price,[6] was able to calculate the value of A in (8.1) for the oxidation of copper assuming the type of mechanism shown in Fig. 8.2, calculated values agreeing well with observed values. Oxide films possessing poor electronic and/or poor ionic conduction will grow slowly and will therefore at least be semi-protective. Aluminium forms a film with very few lattice defects and therefore the oxide which forms initially prevents further attack—but at temperatures above 400°C the concentration of defects in alumina increases and growth becomes parabolic. Cr_2O_3 has poor ionic conductance even though electronic conductance is high—this is the reason for the use of high concentrations of chromium in heat-resisting steels. The more stable oxide of chromium forms a protective layer at the metal/oxide interface.

Departures from parabolic growth often occur, and we will close this short discussion of this important group of heterogeneous reactions by considering briefly the factors causing such departures. Further reading on the subject of "dry" corrosion will be provided by reference to the texts in ref. 7. In the early stages of film growth at high temperatures (e.g. iron at 1000°C), the rate of diffusion may be faster than the rate of supply of oxygen from the gaseous atmosphere, which becomes rate-controlling. Under these conditions, the rate will be independent of time and a linear relationship between y and t will be obtained,

$$y = Ct + D, \tag{8.3}$$

where C and D are constants. Eventually, the film will become thick enough for diffusion to be rate-controlling, and a parabolic relationship (8.2) will be obtained (Fig. 8.5). If breaks in a film occur, perhaps due to strains set up during its formation, direct access of gas to metal may be provided and again linear growth has been observed, for example with copper at

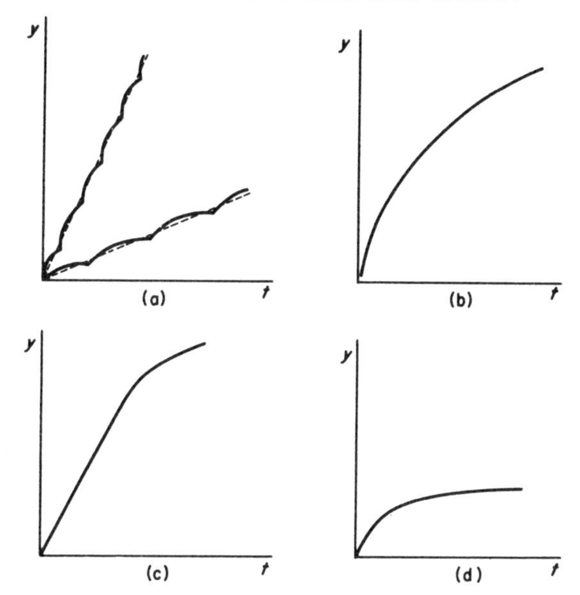

FIG. 8.5. Various rates of growth of oxide films: y = thickness of film, t = time. (a) Linear growth rate produced by intermittent fracture of film growing parabolically. (b) Parabolic growth rate. (c) Linear growth rate in early stages of film growth followed by parabolic growth in the later stages. (d) Logarithmic growth rate.

500°C when the scale is brittle. At higher temperatures the scale is more easily deformed under strain and parabolic growth of copper oxide is observed.[8] The apparent linear nature of the growth when fracturing occurs may in fact be a series of short periods of parabolic growth punctuated by fracture and an acceleration in growth rate (Fig. 8.5). Linear growth rates have been observed with calcium.[8]

At low temperatures, where the thermal agitation is slight and ionic movement would be difficult, or with films such as alumina in which the lack of lattice defects and poor electronic conduction cause a rapid decrease in the rate of growth as the film thickens (Fig. 8.5d), a logarithmic relationship is observed, for example

$$y = E \log (Ft + G), \tag{8.4}$$

where E, F and G are constants. Another cause of logarithmic growth is the coalescence of vacancies to form voids, which, in the case of the cation-deficient p-type semiconducting oxide where the vacancies diffuse inwards (Fig. 8.2), tend to form near the metal/oxide interface. (Voids and cavities of this type have been found in wüstite films.) This restricts the area of film across which mass transport can take place and as these voids increase in size, further restriction reduces the rate of film growth, and gives a logarithmic relationship

$$W = H \ln (Jt), \tag{8.5}$$

where W is the weight increase per unit area and H and J are constants. Other mechanisms have been proposed which suggest logarithmic growth rates and it is often difficult to distinguish between them in accounting for a particular oxide film's growth rate. Evans[9] presents a critical discussion of these mechanisms, and may be consulted in addition to the references under (7).

8.3. The Tendency to Corrode in Aqueous Media

A metallic structure in contact with an aqueous solution is an electrode, and its tendency to dissolve in the aqueous electrolyte will be measured by its electrode potential (Section 5.6). The higher the electrode potential, the lower the driving force behind the process of solution. If the metal is in electronic contact with another substance of higher electrode potential than its own, forming a galvanic cell, the metal will become the anode of the cell and will tend to dissolve anodically, eqn. (5.25)

$$M_1 = M_1^{z_1+} + z_1\epsilon^-.$$

The driving force behind this anodic dissolution is the e.m.f.

of the galvanic cell, the difference between the electrode potentials of the component electrodes (5.28),

$$E_{\text{cell}} = E_2 - E_1.$$

Cells of this type—such as an iron bar bolted to the inside of a water-filled copper tank—lead to *galvanic corrosion* which is the gradual anodic dissolution of the metal with the lower electrode potential—in this case, the iron bar.

In addition to the effect of an applied potential on the tendency of a metal to dissolve anodically (the potential may be applied by contact with an electrode of high potential or by an "impressed" e.m.f. supplied by some external source such as a battery) the pH of the aqueous solution has a profound effect on the products of the anodic dissolution. The combined effect of the applied potential and pH of the aqueous solution on the products of corrosion has been well summarized in *Pourbaix diagrams*—devised by M. Pourbaix[10] and involving plots of the equilibrium values of applied potential and pH for the various equilibria in the metal–water system. The electrode potential of any electrode M/M^{z+}_{aq} depends on the activity of metal ions in the electrolyte [Nernst equation, (5.23)]

$$E = E^{\circ} + \frac{RT}{zF} \ln a_{M^{z+}},$$

so that equilibria which depend on the applied potential will also depend on $a_{M^{z+}}$. Pourbaix diagrams may show lines representing equilibria for various values of $a_{M^{z+}}$, but it is sometimes convenient to choose an arbitrary value for activity of any soluble ionic species in the system, and this is often fixed as a concentration of 10^{-6} mole/l., which is considered fairly representative of typical corrosion conditions. The temperature must also be constant for the construction of these diagrams, and this is usually fixed at 25°C. Pourbaix has now published a collection of potential–pH diagrams for individual metals[11] and Fig. 8.6, the Fe–H_2O diagram, is adapted from that book.

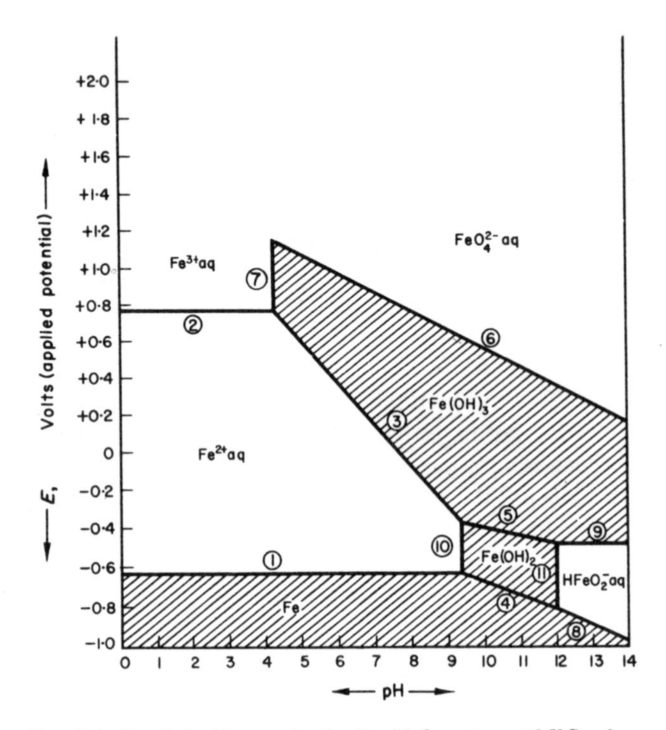

FIG. 8.6. Pourbaix diagram for the Fe–H_2O system at 25°C, where the concentration of all ions in aqueous solution is 10^{-6} mole/l. (After Pourbaix, M.[11])

Referring to Fig. 8.6, which may be interpreted in a similar manner to Fig. 7.5 (Section 7.5) using E (applied potential) and pH as variables rather than $\log p_{SO_2}$ and $\log p_{O_2}$, horizontal lines (1, 2, 9) represent equilibria controlled by E but not pH, for example

1. $Fe^{2+}_{aq} + 2\epsilon^- = Fe$,

which is an electrode reaction whose electrode potential is given by (5.23)

$$E_{Fe} = E^{o}_{Fe} + \frac{RT}{2F} \ln a_{Fe}{}^{2+}.$$

$a_{Fe^{2+}}$ is fixed at 10^{-6} and the value of E on line 1 is in fact the electrode potential of iron in this solution. If the applied potential is above this value, the iron dissolves anodically—for example if it is connected to copper whose electrode potential is higher than line 1. If the applied potential is below line 1, the iron remains immune—for example, if it is connected to zinc, which itself will dissolve anodically in preference to iron. Line 2 represents the equilibrium

$$Fe^{3+}_{aq} + \epsilon^- = Fe^{2+}_{aq},$$

and the corresponding redox potential is given by (5.33). $a_{Fe^{2+}} = a_{Fe^{3+}} = 10^{-6}$ so that E for line 2 is E^o, the standard electrode potential for the redox electrode (0·77 V at 25°C). Vertical lines represent equilibria which involve hydrogen ions but not a transfer of electrons so that they are independent of E. An example of a vertical line is line 10, which represents the equilibrium

$$Fe^{2+}_{aq} + 2H_2O = Fe(OH)_2 + 2H^+,$$

which is the hydrolysis of ferrous ions to form the insoluble hydroxide. This hydroxide is probably a hydrated oxide of iron (FeO,H_2O). Diagonal lines represent equilibria involving both hydrogen ions and electron transfer, for example line 4

$$2H^+ + Fe(OH)_2 + 2\epsilon^- = Fe + 2H_2O,$$

where
$$E = E^{\circ}_{Fe/Fe(OH)_2} + \frac{RT}{F}\ln a_{H^+},$$

$$= E^{\circ}_{Fe/Fe(OH)_2} - 0{\cdot}059\,\text{pH},$$

at 25°C, using the definition of pH as $-\log_{10}a_{H^+}$. Iron held in a solution of pH 11 and raised to a potential of −0·5 V will not corrode—instead it will be *passivated* because an insoluble layer of $Fe(OH)_2$ will form on its surface. A further increase in potential to +0·2 V would still cause passivation by the formation of $Fe(OH)_3$ (probably hydrated Fe_2O_3) and there would only be a danger of corrosion above line 6. Passivation depends on the physical nature of the passivating film, and corrosion will only cease if the film is completely impervious—a porous film will slow down corrosion but not stop it altogether. This phenomenon is kinetic in its effects and will be considered again in the next section. Highly alkaline solutions at low potentials can cause corrosion of iron by the formation of the soluble bihypoferrite,

$$8.\quad HFeO^-_{2aq} + 3H^+_{aq} + 2\epsilon^- = Fe + 2H_2O,$$

which again depends on E and pH, and slopes from left to right. The shaded regions of Fig. 8.6 can be considered to represent conditions in which iron is either immune or exhibits passivity.

Pourbaix diagrams illustrate the tendency of a metal to corrode in aqueous solution at the temperature and of the composition chosen for the diagram, but they do no more than this because they are constructed from *equilibrium* measurements. Corrosion *rates* are not indicated and will be considered in the next section.

The *cathode reaction* in a corrosion cell could be (5.24),

$$M_2^{z_2+} + z_2\epsilon^- = M_2,$$

but the concentration of metal ions in a typical corrosive electrolyte will be small and the reaction usually taking place is either

$$H^+_{aq} + \epsilon^- = \tfrac{1}{2}H_2 \tag{8.6}$$

or

$$\tfrac{1}{2}O_2 + H_2O + 2\epsilon^- = 2(OH^-)_{aq}. \tag{8.7}$$

In neutral (pH approximately 7) solutions where the solution contains dissolved oxygen due to aeration, (8.7) will tend to predominate because the hydrogen ion concentration will be low, so that a galvanic corrosion cell under these conditions will have an overall reaction

$$M_1 + O_2 + H_2O = M^{2+}_{1aq} + 2(OH^-)_{aq},$$

where M_1 corrodes, forming the divalent ion M_1^{2+}. The cell e.m.f. will be the difference between that for the oxygen electrode, $M_2|O_{2aq}$, and the $M_1|M_1^{2+}$ electrode. M_2 is the cathode but, providing its electrode potential is higher than that for M_1, it will not corrode and the reversible cell e.m.f. will be independent of the nature of M_2. (Polarization will, however, depend on the electrode material.) For example, in the cell $Fe|Fe^{2+}_{aq}|O_{2aq}|Cu$, the cell e.m.f. will be independent of the fact that the metal cathode is copper, and will depend on the oxygen electrode potential,

$$E_{O_2} = E^0_{O_2} + \frac{RT}{2F} \ln \frac{(p_{O_2})^{1/2}}{a^2_{(OH^-)}}. \tag{8.8}$$

If p_{O_2} varies at different points on a metallic structure, a *differential aeration* cell can be set up—see Section 5.10.

The oxide film on the surface of a metal plays an important role in the corrosion process, and it is not necessary for a second metal to be present for corrosion to take place. The oxide film—which will be present on metals forming stable oxides (Section 8.2)—can act as the cathode in a galvanic cell formed with the aqueous electrolyte and the metal itself, laid bare to the electrolyte at breaks and weaknesses in the oxide film (Fig. 8.7). The oxide (which might be partially

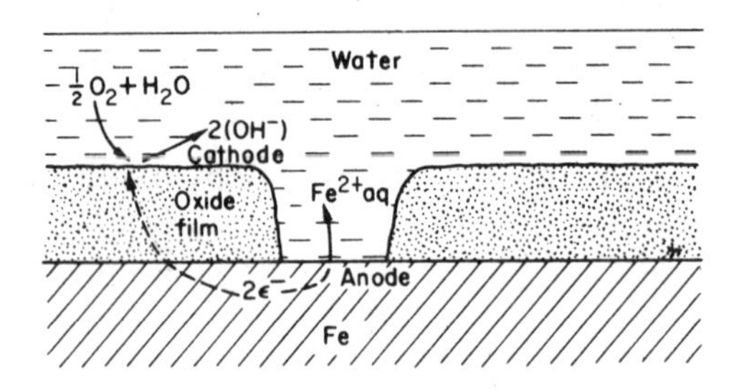

FIG. 8.7. Corrosion of iron in aerated neutral aqueous solution at breaks in the oxide film.

hydrated in aqueous solution) must be an electronic conductor to supply the electrons for the cathode reaction (8.7), and the metal will corrode at the anodic points—with the possibility that the oxide film may be undermined by the solution of the metal, eventually breaking away to lay bare a greater area of anodic metal. In the case of iron, the hydroxyl ions produced by the cathode reaction (8.7) will react with the metal ions (oxidized from ferrous to ferric by the presence of oxygen in the solution) to precipitate a porous "rust"—hydrated ferric oxide (Fe_2O_3, xH_2O)—near the anodic region if the oxygen content of the water is high. (Precipitation occurs at a greater distance if the oxygen content is low.)

Anode reaction: $Fe \rightarrow Fe^{2+}_{aq} + 2\epsilon^-$.

Cathode reaction: $\frac{1}{2}O_2 + H_2O + 2\epsilon^- \rightarrow 2(OH^-)_{aq}$.

The anodic and cathodic products then react

$$Fe^{2+}_{aq} \rightarrow Fe^{3+}_{aq} + 3(OH^-)_{aq} \rightarrow Fe_2O_3, xH_2O.$$

The hydrated oxide is too porous when formed under these conditions to block the anode area, and the corrosion process will tend to continue under the steadily growing volume of rust which forms as the corrosion product. The presence of dissolved salts such as sodium chloride will increase the conductance of the electrolyte and will therefore increase the corrosion current and the rate of corrosion.

If the pH of the solution is low—as in the case of dilute solutions of the mineral acids—the oxide film will dissolve in the acid and corrosion of bare metal will take place. The identity of the anode and cathode in such a process, where a very pure metal is involved, has been the subject of some speculation. The surface of a metal has many imperfections—including grain boundaries, locally strained regions and the ends of dislocations—which are chemically active compared with the remainder of the surface. These imperfections may provide anodic sites and the perfect surface will act as the cathode. In an impure metal, impurity atoms or non-metallic inclusions with higher electrode potentials than the metal will act as cathodes. The cathode reaction will be mainly (8.6) because the hydrogen ion concentration is high in acid solutions, although (8.7) will also take place if the solution is aerated. It must be remembered that (8.6) can only take place if the metal's electrode potential is *below* that of the hydrogen electrode.

With oxidizing acids, alternative cathodic reactions will be possible, such as the following multi-stage reaction in nitric acid:[12]

$$H^+ + NO^-{}_3 = HNO_3$$

$$HNO_3 + HNO_2 = N_2O_4 + H_2O$$

$$N_2O_4 = 2NO_2$$

$$2NO_2 + 2\epsilon^- = 2NO_2^-$$

$$2NO_2^- + 2H^+ = 2HNO_2,$$

giving an overall reaction

$$3H^+ + NO_3^- + 2\epsilon^- = HNO_2 + H_2O. \qquad (8.9)$$

The electrode potential of the electrode whose reaction (8.9) represents is high ($E^o = +0{\cdot}93$ V) so that metals such as copper, which lie above hydrogen in the Standard Electrode Potential Series ($E^0_{Cu} = +0{\cdot}34$ V), will corrode in nitric acid whereas they will not corrode in dilute hydrochloric acid. With metals such as iron and zinc, (8.6) can also take place and the hydrogen produced reacts with the products of (8.9) to give ammonium ions $(NH_4)^+$ and other reduction products of nitric acid.

In *alkaline* solutions, we saw from the Pourbaix diagram (Fig. 8.6) that iron tends to dissolve anodically as the bihypo-ferrite, $HFeO_2^-$. Aluminium and zinc react similarly, forming aluminate anions, AlO_2^-, and zincate anions, ZnO_2^{2-}, respectively. For example, the anode reaction for the corrosion of zinc in a solution of pH 12–14 is

$$Zn + 4(OH^-)_{aq} = ZnO^{2-}_{2aq} + 2H_2O + 2\epsilon^-.$$

The cathode reaction will be

$$H_2O + \epsilon^- = OH^-_{aq} + \tfrac{1}{2}H_2, \qquad (8.10)$$

and hydrogen will be evolved. Equation (8.10) is more probable here than (8.6) because the hydrogen ion concentration in alkaline solution will be very low. Nickel is used as a container for alkaline solutions because it tends to form a passivating layer of $Ni(OH)_2$ in strongly alkaline solutions.

If a metal is subjected to *stresses*, its tendency to corrode and the mode of corrosion may be altered significantly. This topic would require a treatment beyond the scope of this book and the reader is advised to consult refs. 1–4 and two review papers[13] for a survey of the theory and experimental work carried out on corrosion of stressed metals.

8.4. The Rate of Corrosion in Aqueous Media

Electrode potentials and cell e.m.f.'s discussed in the last section were measured under conditions as near thermodynamically reversible as possible – that is virtually no current being passed by the cell. Corrosion is a dynamic process, and depends on the passage of a current in the corrosion cell. Electrode potentials and reversible cell e.m.f.'s are therefore only a measure of the tendency for a cell to work and do not reflect working conditions. If for some reason the rate of a corrosion process is slow, this is really more important to the engineer than the fact that the reversible e.m.f. of the corrosion cell is large. In Section 5.12 it was shown that as soon as a galvanic cell passes a current, its e.m.f. is reduced because the apparent electrode potential of the cathode falls and that of the anode rises – causing a decrease in the difference between them. This change in an electrode potential is due to *polarization*.

The effects of polarization can be demonstrated for a galvanic corrosion cell by plotting the electrode potential

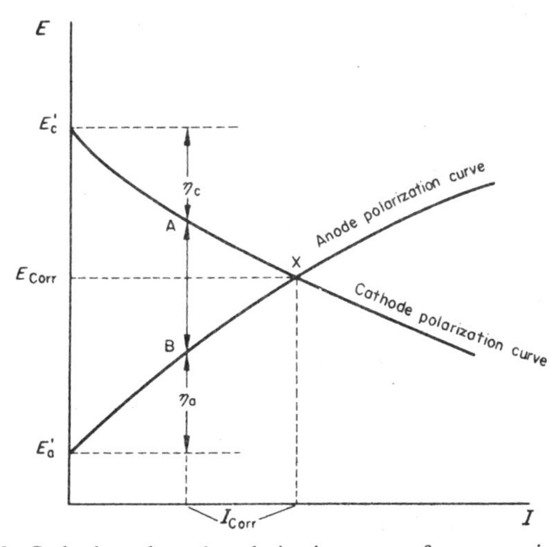

FIG. 8.8. Cathode and anode polarization curves for a corrosion cell.

of the cathode and anode against the current passed by the cell (Fig. 8.8). E is the potential of the electrode (E_c for the cathode, E_a for the anode) measured using an apparatus such as that in Fig. 5.23. As the current I passed by the cell increases, E_c falls from the reversible cathode potential E_c' and E_a rises from the reversible anode potential E_a'. Eventually the two curves cross at the point X, and the current at this point is the maximum current delivered by the cell. This can be demonstrated by reference to Fig. 8.9, which represents a

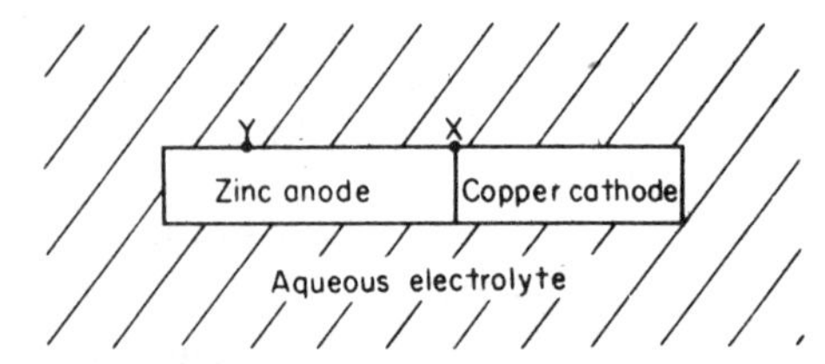

FIG. 8.9. Composite bar of copper and zinc immersed in an aqueous electrolyte.

composite bar of copper and zinc immersed in an aqueous electrolyte. The corrosion cell is made up of the copper cathode, zinc anode, aqueous electrolyte and electronically conducting joint between the two metals. At the point X, the electrolyte resistance between the two electrodes is a minimum and the corrosion of the zinc, whose rate is proportional to the corrosion current (I_{corr}), occurs more rapidly than at the point Y, where the electrolyte resistance R between anode and cathode becomes significant. AB in Fig. 8.8 is equal to $I_{corr}R$, or $(E_c - E_a)$. Note that, at this point,

$$E_c' - E_a' = (E_c - E_a) + |\eta_c| + |\eta_a|, \qquad (8.11)$$

where η_c and η_a are the cathode and anode overpotentials respectively. If the value of R for point Y is negligible, then the rate of corrosion will be the same for X and Y, and the value of I_{corr} at X is usually called the *corrosion current*. At

X, the anode and cathode will both be at the same potential, E_{corr}—called the *corrosion potential.* Any factor which increases the polarization at either the anode or the cathode will decrease I_{corr} and therefore reduce the rate of anodic dissolution.

Often, polarization at one of the electrodes dominates the corrosion process, and we will discuss some examples of this phenomenon. In aerated neutral solutions, the cathode reaction (8.7) requires the consumption of oxygen and at high currents the supply of oxygen becomes rate-controlling. This causes *concentration polarization* to become more significant at high corrosion currents than the *activation polarization* which already makes the polarization of the cathode reaction greater than that at the anode (Fig. 8.10). Curve A is for stagnant

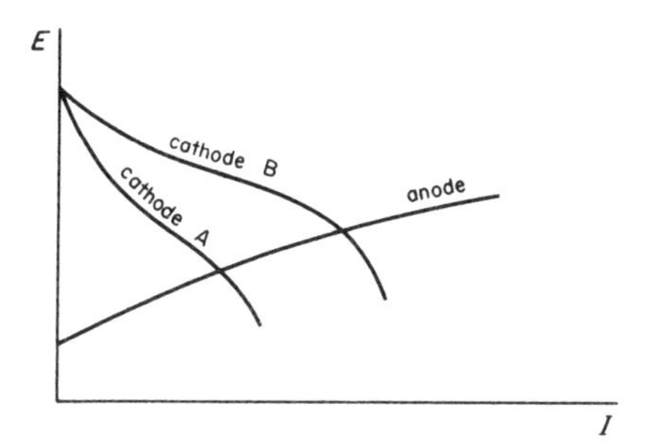

FIG. 8.10. Oxygen supply to the cathode causing polarization at the cathode to become rate-controlling.

water conditions and curve B for rapidly moving water conditions in which the oxygen supply is greater. Clearly, if no other factors are significant, the corrosion rate will be the greater in case B because the value of I_{corr} will be higher.

If the electrolyte is an acid solution, the cathode reaction will be (8.6), and the hydrogen overvoltage may be rate-controlling especially with certain metals on which the hydrogen overvoltage is high (Fig. 8.11). The hydrogen overvoltage on copper is higher than that on platinum so that iron

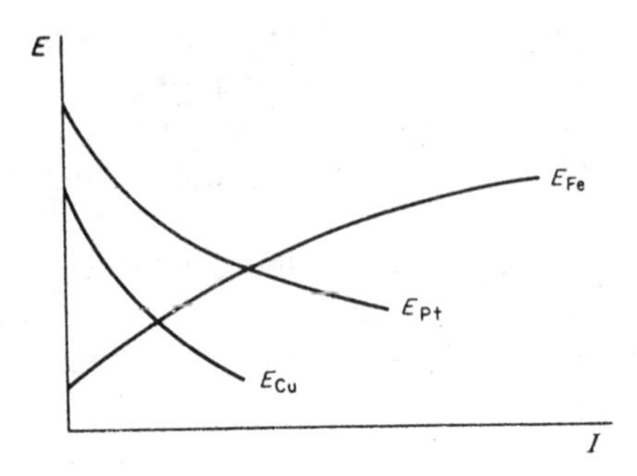

FIG. 8.11. Cathodic control due to hydrogen overvoltage, in corrosion cells with an iron anode and platinum and copper cathodes.

would corrode more rapidly in acid solution in contact with platinum than it would with a copper cathode. Zinc has a high hydrogen overvoltage so that pure zinc corrodes much more slowly in mineral acids than zinc containing impurities such as copper which lower the hydrogen overvoltage on zinc.

An example of anodic control of a corrosion process is *passivation* in which the metal reacting anodically does not go into solution but forms a layer of oxide or hydroxide over the surface of the metal which introduces a form of *resistance overpotential* by limiting access of electrolyte to the metal surface (Fig. 8.12). Aluminium, stainless steel and titanium are sufficiently reactive to form this passivating film in aerated water, but iron requires a higher pH than a neutral solution (Fig. 8.6) or the presence of an oxidizing agent such as concentrated nitric acid,

$$Fe = Fe^{2+} + 2\epsilon^-$$

$$3H^+ + NO_3^- + 2\epsilon^- = HNO_2 + H_2O \qquad (8.9)$$

$$Fe^{2+} + \tfrac{3}{2}NO_2^- + 6H^+ + \epsilon^- = \underset{\text{passivating oxide}}{Fe_2O_3} + \tfrac{3}{2}(NH_4)^+.$$

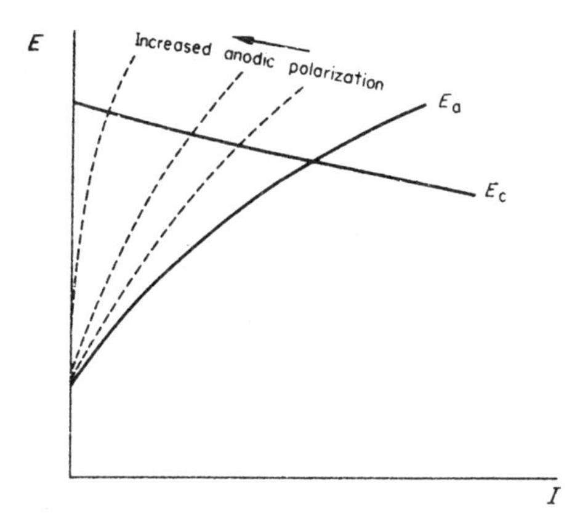

FIG. 8.12. Increased anodic polarization causing a reduction in corrosion current. Passivation is the condition in which anodic polarization is so severe that virtually no current flows.

The effect of passivation of reactive metals like titanium and aluminium in aerated water is to make them behave as if they had higher electrode potentials than suggested by the Standard Electrode Potential Series – and this has led to the construction of "galvanic series" – lists of electrode potentials of metals and alloys in particular corrosive conditions. For example, titanium comes at the top of a galvanic series using sea water as the electrolyte rather than near the bottom as in Table 5.1, p. 196. De-aeration of a solution can cause breakdown of the passivating film of chromium oxide on stainless steel which is then no longer "stainless". The presence of the chloride ion, Cl^-, in the electrolyte seems to cause breakdown of the oxide film on stainless steel, producing localized "pitting" corrosion.

The phenomenon of passivation has been studied by means of *potentiostatic* measurements in which the potential applied to a metal in an aqueous electrolyte can be controlled and the current flowing will adjust itself to maintain that potential. The potential is increased and the corresponding current density is measured. If E, the applied potential, is plotted against $\log i$, where i is the current density at the metal surface, a curve of the type shown in Fig. 8.13 will be obtained. At A, the potential is high enough for anodic dissolution of the metal to begin (see Pourbaix diagram, Fig. 8.6). As E rises, log i increases giving a linear Tafel relationship (5.37) until the critical current

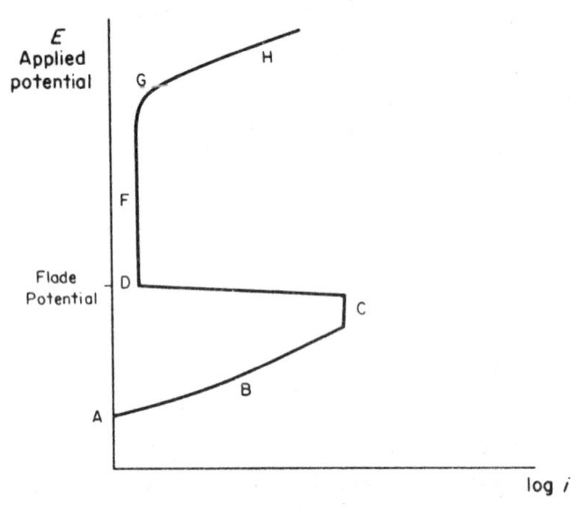

FIG. 8.13. Potentiostatic curve showing passivation of a metal.

density i_{crit} is reached at C. Passivation begins as the oxide film forms and at D, called the *Flade potential*, a minimum current density is reached and the metal is passivated. Increasing potential (F) does not increase the current density, and dissolution of the metal can only occur through the oxide film. At higher potentials (G), the potential for the oxygen electrode is reached and oxygen is evolved as E increases along H, giving an increase in current density. Oxygen is evolved by the reverse of reaction (8.7),

$$2(OH^-)_{aq} = \tfrac{1}{2}O_2 + H_2O + 2\epsilon^-.$$

The Flade potential is roughly related to the reversible potential of the electrode consisting of the metal and its oxide, M/MO, but is not always identical with this potential. Metals like chromium and titanium have low Flade potentials and are easily passivated, iron has a higher Flade potential and must be raised above a potential of + 0·6 V by the cathodic reaction of nitric acid (8.9) to become passivated.

Current density at an electrode has an important effect on polarization (Section 5.12) and is also a measure of the rate of penetration of a metal by corrosion. The higher the anode current density, the greater the weight loss over a given area, and therefore the combination of a large cathode and a small anode can cause severe corrosion rates. An example of this is aluminium—whose protective oxide film may become weakened at isolated points. These points will be anodic to the rest of the oxide surface and severe *pitting* can occur if the corrosive environment prevents passivation of the anode. Increasing current density tends to increase polarization and it is possible that the current density at a small anode can become so high that polarization stifles the corrosion reaction. A cathodically controlled corrosion process will cause very severe corrosion if the cathode is large compared with the anode—for example, in aerated neutral solution, oxygen supply to the cathode will be favoured by a large cathode.

8.5. The Prevention of Corrosion

The only way to prevent corrosion may be to choose a metal or alloy which resists corrosion—such as nickel for alkaline solutions, tantalum for most environments (at a high price!), magnesium for hydrogen fluoride (it forms an insoluble passivating film of magnesium fluoride), stainless steels containing large amounts of chromium for aerated solutions in the absence of chloride ions. Alloying for corrosion resistance can be carried out to some extent without impairing mechanical properties—for example, a small amount of aluminium added to a brass causes a tight oxide film of aluminium to form protectively on the surface of the brass. Corrosion-resistant metals or alloys may be either too expensive or of the wrong mechanical properties for many applications, but *good design* can go a long way towards prevention of corrosion – for example, the avoidance of crevices in which differential aeration cells might be set up and of electronically conducting joints between dissimilar metals immersed in water.

Cathodic protection is the application of the information from a suitable Pourbaix diagram (Fig. 8.6). If the potential of a metal can be *lowered* sufficiently, it will become immune. This can be achieved by placing it in contact with a second metal of lower electrode potential, which will dissolve anodically in preference to the metallic structure being protected. An example of this form of sacrificial protection is galvanized iron – mild steel sheet coated with zinc. The zinc may prevent contact between the steel and the aerated aqueous electrolyte mechanically, but at breaks in the zinc coating, the iron will be the cathode in a galvanic cell – the zinc corroding preferentially. Only where large areas of iron are exposed will corrosion of the iron occur. A second method of applying cathodic protection is by using a d.c. source to apply the potential, for example to protect buried pipelines (Fig. 8.14).

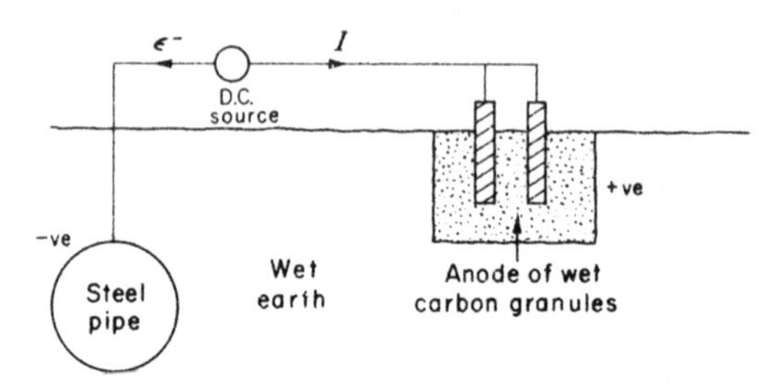

FIG. 8.14. Cathodic protection of buried pipelines.

The anodes are sunk in beds along the length of the pipeline and currents of several mA per square foot of bare metal are required, so that it is usual to coat the pipeline with some form of impervious material (bitumen) to reduce the area of exposed metal.

Inhibitors may be used under certain circumstances,

although the addition of chemicals to very large volumes of water may be impracticable and in certain items of chemical plant may interfere with the chemical process. *Cathodic inhibitors* in neutral solution suppress reaction (8.7) at the cathode in a corrosion cell,

$$\tfrac{1}{2}O_2 + H_2O + 2\epsilon^- = 2(OH)^-_{aq}.$$

This process can be an encouragement of concentration polarization at the cathode by mopping up the oxygen with something like sodium sulphite,

$$Na_2SO_3 + \tfrac{1}{2}O_2 = Na_2SO_4,$$

or by de-aeration of large volumes of water by passing it over scrap iron, which absorbs oxygen as it rusts. Alternatively, resistance polarization can be caused by blanketing the cathode with substances such as hydroxide precipitates which are impermeable to both oxygen and electrons. Use is made of the presence of Mg^{2+} and Ca^{2+} in hard waters to precipitate a protective layer of $Mg(OH)_2$ or $Ca(OH_2)$. *Anodic inhibitors* suppress the anodic dissolution reaction by causing resistance polarization of the anode—for example with iron, the Fe^{2+} ions may be precipitated as an impervious layer as soon as they are produced by the dissolution at the anode and have been oxidized to Fe^{3+}. This can be achieved by the addition of chromates or phosphates, which form insoluble iron salts at the anode. The inhibitor precipitates with the iron but the main part of the sealing of the anodic areas is by iron oxide which precipitates as a compact layer rather than porous "rust". The amount of inhibitor is critical and if less than the critical amount is present, a few anodic sites will remain and severe pitting will result. Another form of anodic inhibitor, particularly useful in acid solutions, is an organic substance which is soluble in water and has a polar group which causes it to be strongly adsorbed on the surface of a metal (Section 6.4). As soon as bare metal appears at an anodic site, the inhibitor becomes strongly adsorbed on the metal and prevents further attack because a non-polar organic group is presented to the electrolyte rather than the metal surface. Sulphonic acids and amines are particularly effective in installations carrying oil and water. The organic inhibitor probably becomes adsorbed on the surface of the metal and its non-polar group dissolves easily in oil—so that the oil covers the metal surface rather than the water.

Protective coatings can be used to prevent or reduce corrosion. These may be organic coatings such as bitumen, greases, or plastics—all being purely mechanical in their exclusion of the corrosive medium. Paints are somewhat porous and appear to contain anodic inhibitors. Metallic coatings can be mechanically protective (tinplate—a coating of tin on mild steel sheet—tin, being cathodic to the steel, can only be mechanical in its protective action) or sacrificially protective (zinc on mild steel). The metallic coating can be applied by dipping the metallic structure in liquid metal (zinc, tin), by spraying low melting point metals on the surface (zinc, aluminium), or by cladding—rolling two sheets of metal together (stainless steel on mild steel, silver on copper). Finally, the metallic coating may be deposited electrolytically, and this will be the subject of the last section of this chapter.

8.6. Electrodeposition

The use of electrodeposition in extraction and refining of metals was discussed at the end of the last chapter, and here we will be concerned with *electroplating*[14] of metals from aqueous solutions, although the principles are generally applicable to all electrodeposition processes. Electroplating is often a convenient method of obtaining a metallic coating whose properties are as far as possible adherent, compact, non-porous, even, ductile, and frequently a bright finish. The coating may be very thin, and may be required for its appearance as well as its protective action.

In Section 5.13 it was made apparent that a metal ion will be discharged at a cathode when the potential of the cathode is less than the discharge potential of that ion under the conditions existing in the cell—the discharge potential being the sum of the reversible electrode potential E_r and the cathodic overpotential η_c. E_r will depend on the activity of the metal ions in the electrolyte (5.23),

$$E_r = E_M^0 + \frac{RT}{zF} \ln a_{M^{z+}},$$

and the activity of M^{z+} will be controlled by the ligands in the solvation sheath round the ion—these may be entirely water molecules but could be cyanide ions forming a complex ion, (Section 5.2). η_c will be negative, and will depend on the current density, the mechanism of the deposition process and changes in concentration near the cathode surface, being the sum of activation, concentration and resistance overpotentials, (5.39):

$$\eta_c = \eta_c^A + \eta_c^C + \eta_c^R.$$

The anode used may be non-consumable or of the same metal as that being discharged at the cathode, and consequently consumable. In the former case the electrolyte must be replenished but in the latter this will not be necessary. If the working potential approaches the discharge potential of hydrogen from the electrolyte used, hydrogen will be increasingly discharged at the cathode—with a consequent loss in current efficiency (Section 5.3), because current will be used up in discharging an unwanted species, and with possible damage to the cathode deposit where bubbles have adhered to the deposit and hindered smooth metallic deposition. This becomes increasingly important with metals having negative standard electrode potentials so that metals below zinc in the standard electrode potential series (Table 5.1) must be deposited from non-aqueous solutions, for example aluminium from a solution in ether.

pH control can be important in plating electrolytes, and the Pourbaix diagram for nickel (Fig. 8.15) demonstrates the possibility of $Ni(OH)_2$ precipitating from neutral solutions

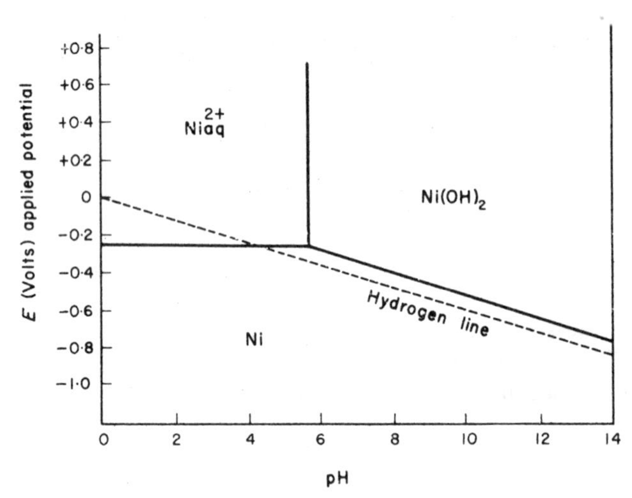

FIG. 8.15. Portion of Pourbaix diagram for Ni–H_2O at 25°C. Broken line shows the potential below which hydrogen discharges from solution at the value of pH indicated. (Concentration of all ions present in aqueous solution is 1 mole/l.) (After Pourbaix, M.[11])

and becoming caught up in the nickel deposit. An acid solution is required, but the pH must not be allowed to fall too low, otherwise hydrogen would tend to discharge more easily—the line for the discharge of hydrogen

$$H^+ + \epsilon^- = \tfrac{1}{2}H_{2_G}$$

being of negative slope. A buffer is used in the electrolyte to control pH in these circumstances—for example, boric acid in a nickel sulphate/nickel chloride electrolyte to maintain a pH between 4 and 5.

The aim in electroplating is to produce a deposit of equal thickness over the whole article being plated, but if a complex shape is used as the cathode, the distance between anode and cathode will vary at different points on the cathode surface—giving different electrolyte resistance for each point. The lower the electrolyte resistance, the higher the current and hence the greater thickness of metal deposited in a given time—this will tend to produce an uneven deposit. If polarization can be increased at positions nearest the anode and if, as far as possible, anodes are arranged in shapes and positions to take account of the cathode shape, an approximately even deposit may be obtained. An electroplating bath which produces a good deposit at points comparatively remote from the anode is said to have good *throwing power*. The Haring cell (Fig. 8.16) is used to measure throwing power (*T.P.*) for a particular electrolyte/current density/cathode potential combination. The cell is used for a definite time under conditions similar to those used in practice, and the ratio of metal deposited at cathode 1 to that at cathode 2 (M) is related to the ratio of l_2 to l_1 (L), the respective distances between the cathodes and the anode. Various formulae have been proposed, but the British Standard formula is that due to Field,

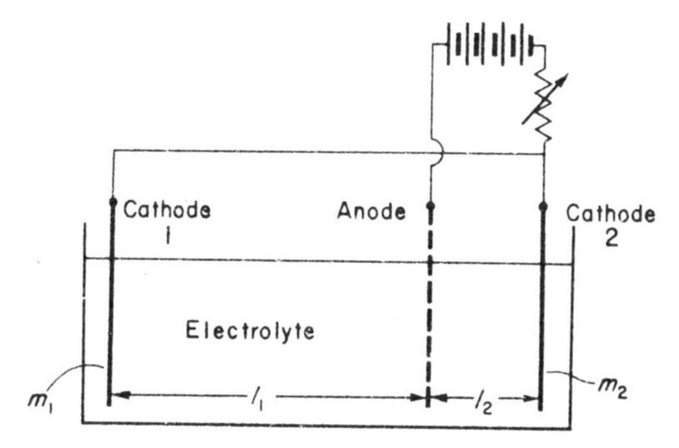

FIG. 8.16. The Haring cell for measurement of the throwing power of an electroplating bath. m_1 and m_2 are the weights of metal deposited in a given time at the respective cathodes.

$$T.P. = \frac{L-M}{L+M-2} \times 100. \qquad (8.12)$$

If $L = M$, $T.P. = 0$; if $M = 1$, $T.P. = +100\%$; if no metal plates out at cathode 2, $T.P. = -100\%$. Hoar and Agar[15] deduced that, where the cathode current efficiencies of the two electrodes are the same and κ is the specific conductance of the electrolyte, the ratio of the cathode current densities is given by

$$\frac{i_1}{i_2} = \frac{l_2 - \kappa(\Delta E_c/\Delta i)}{l_1 - \kappa(\Delta E_c/\Delta i)} = M. \qquad (8.13)$$

$\Delta E_c/\Delta i$ is the average slope of the curve plotted of the cathode potential E_c against i, the current density, between the values of i_1 and i_2 (Fig. 8.17). If κ is high, M will be approximately

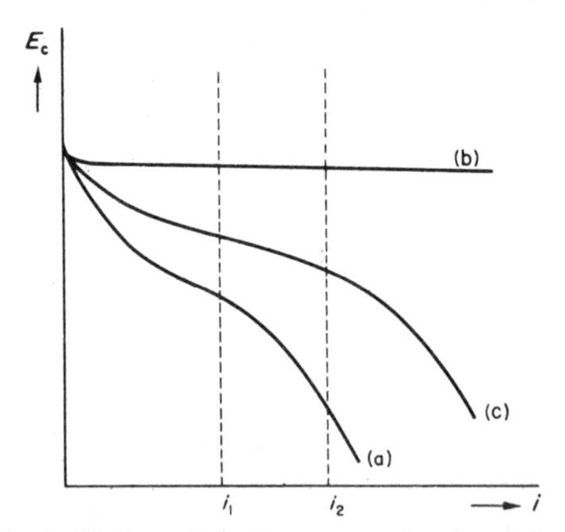

FIG. 8.17. Cathodic polarization curves for three electroplating baths working in the same range of current densities. (a) has a high average slope $\Delta E_c/\Delta i$ between i_1 and i_2—giving a high throwing power. (b) $\Delta E_c/\Delta_i = 0$, giving zero throwing power. (c) has intermediate throwing power.

unity, giving good throwing power. If the slope of the polarization curve, $\Delta E_c/\Delta i$, is high, again $T.P.$ will be high. A large value of l_2 and l_1 (that is a large distance between anode and cathode) will give poor throwing power. It is consequently better to work with high cathodic polarization—which is favoured by low temperatures and the absence of stirring—and this is not always in accord with a desire for a high production rate. Polarization is higher in the case of baths in which the metal is present as a complex anion than those in which the metal is present in the electrolyte as a cation. Consequently, the throwing power of cyanide baths should be higher than, say, sulphate baths.

The surface finish of the deposit is usually of importance and study of the mode of crystallization of the metal on the cathode surface has shown that a well-defined, coarsely crystalline surface—giving a rough surface finish—is favoured by low current densities and low cathodic overpotentials. Deposition tends to occur on definite crystallographic planes, but if the current density and cathodic polarization are increased, the rate of nucleation increases, giving a fine grain size, and the deposition is more random—not so closely associated with definite crystallographic features. The result is a surface which tends to reflect light more perfectly, with less scattering. Bright electrodeposition is assisted by the addition of certain organic compounds possessing polar or charged groups which allow these compounds to become adsorbed at active sites on the metal surface—causing the metal to deposit at other sites which would not normally be the positions at which growth nuclei form. The presence of the adsorbed molecules increases cathodic polarization and the combination of this and the randomizing effect of the addition agent gives a bright, fine-grained deposit with imperfections less than 50 Å high in the surface—small enough to give specular reflection. Sometimes the brightener may be caught in the deposit—causing embrittlement. Brighteners of this type include glue, gelatine, carbohydrates, and sulphur-containing compounds such as thiourea, thiocarbamates. Another form of brightener has a smaller molecule and is used in higher concentrations—for example, formaldehyde

(HCHO), cyanide, fluoride ions and ammonia. These species are not adsorbed strongly on the metal surface, but there is a high rate of exchange – leaving a large proportion of the surface covered at any moment. The continual change of adsorption site causes random metal deposition and increases polarization – producing a bright finish.

Complexing agents are used (especially the cyanide ion) to replace water in the solvation sheath around the metal ion and thereby to alter the discharge potential of the metal ion and the cathodic polarization. The result is a better surface finish of the deposit and a change in the potential at which ions discharge. If a metal with a high electrode potential such as copper is to be plated on steel – whose potential is considerably lower (Table 5.1) – difficulty will be experienced because the iron will displace the copper in solution by a cementation process (Section 5.9) giving a porous non-adherent coating of copper as soon as the steel is dipped in the electrolyte. To prevent this a "copper strike" bath is used – a thin flash of copper being deposited initially from a cyanide bath in which the activity of copper is lowered sufficiently to prevent the cementation reaction. The plating of alloys becomes possible if the noble metal in the alloy can be complexed sufficiently to lower its discharge potential to that of the other metal. For example, 70:30 brass can be deposited by lowering the activity of copper ions as cyano-complexes sufficiently to allow deposition of 30% zinc concurrently with 70% copper.

References

1. EVANS, U. R. *The Corrosion and Oxidation of Metals*, Edward Arnold, London, 1960.
2. UHLIG, H. H. *Corrosion and Corrosion Control*, John Wiley, New York, 1963.
3. SHREIR, L. L. (Ed.) *Corrosion*, V **1** and **2**, Newnes, London, 1963.
4. WEST, J. M. *Electrodeposition and Corrosion Processes*, Van Nostrand, London, 1965.
5. DARKEN, L. S. and GURRY, R. W. *J. Amer. Chem. Soc.* **68** (1946), 798.
6. WAGNER, C. *Z. Phys. Chem.* **21** (B) (1933), 25: HOAR, T. P. and PRICE, L. E. *Trans. Faraday Soc.* **34** (1938), 867.

7. KUBASCHEWSKI, O. and HOPKINS, B. E. *Oxidation of Metals and Alloys*, Butterworths, London, 1962: HAUFFE, K. *Oxidation of Metals* (Springer, Berlin, 1956). English Translation, Plenum Press, New York, 1965: BÉNARD, J. *Met. Reviews of the Inst. Metals*, **9** (1964), 473.
8. PILLING, R. B. and BEDWORTH, R. E. *J. Inst. Metals*, **29** (1923), 529.
9. Ref. 1, Chapter XX.
10. POURBAIX, M. *The Thermodynamics of Dilute Aqueous Solutions*, Edward Arnold, London, 1949. (Translation.)
11. POURBAIX, M. *Atlas d'équilibres électrochimiques*, Gauthier-Villars, Paris, 1963. (English translation by FRANKLIN, J., Pergamon, London, 1966.)
12. Ref. 4, pp. 55, 56.
13. PARKINS, R. N. Stress-Corrosion Cracking, *Met. Reviews of the Inst. Metals*, **9**, No. 35 (1964), 201: GILBERT, P. T. Corrosion Fatigue. *Met. Reviews*, **1**, No. 10 (1956), 379.
14. OLLARD, E. A. and SMITH, E. B. *Handbook of Industrial Electroplating*, Iliffe, London, 1958: ref. 3, **2**, Section 13.1: ref. 4, Chapter 5.
15. HOAR, T. P. and AGAR, J. N. *Disc. Faraday Soc.*, No. 1 (1947), Butterworths, London, 1961, p. 162.

Index